LIVING IN A CONTAMINATED WORLD

I dedicate this work to the Spokane Tribe of Indians and their Wellpinit and Ford, Washington neighbors, as well as the people of Millersburg and Albany, Oregon. It is an honor not only to tell, but to learn from, their story. Thank you.

Living in a Contaminated World
Community Structures, Environmental Risks and Decision Frameworks

ELLEN OMOHUNDRO

LONDON AND NEW YORK

First published 2004 by Ashgate Publishing

Reissued 2019 by Routledge
2 Park Square, Milton Park, Abingdon, Oxon, OX14 4RN
52 Vanderbilt Avenue, New York, NY 10017

Routledge is an imprint of the Taylor & Francis Group, an informa business

Publisher's Note
The publisher has gone to great lengths to ensure the quality of this reprint but points out that some imperfections in the original copies may be apparent.

Disclaimer
The publisher has made every effort to trace copyright holders and welcomes correspondence from those they have been unable to contact.

A Library of Congress record exists under LC control number:

ISBN 13: 978-0-8153-9028-2 (hbk)
ISBN 13: 978-1-138-35640-5 (pbk)
ISBN 13: 978-1-351-15376-8 (ebk)

Contents

List of Tables		*ix*
List of Figures		*x*
Preface		*xi*
Acknowledgements		*xiv*
List of Abbreviations		*xv*

1   Bringing Community into Environmental Decision-making:
    An Overview of the Theoretical Orientation

    Introduction   1

    Shared History   5

    Community Identity   8

    Control in Local Decisions   12

    Distribution of Power among Local Institutions   15

    Participation in Decisions   17

    Final Thoughts   21

    Notes   22

2   Methods: New Strategies for Old Problems

    Introduction   24

    Case Selection   25

        Teledyne Wah Chang Albany   26

        Dawn Mining Company   26

        Summary   28

    Operational Premises   28

    Data Collection   29

        Historical Documents   29

        Key Informant Interviews   30

        Respondents   31

        Nonrespondents   32

    Data Analysis   32

    Notes   33

3   Defining Environmental Risks: An Overview of the Superfund Process

    Introduction   34

    Preliminary Assessment Petitions   35

    Hazard Ranking System   36

    Public Participation in Superfund Decision-making Activities   38

    Summary   40

4   This Land is Whose Land?
The Role of Shared History and Community Identity
in Environmental Decision-making
    Introduction 41
    Geography 42
    Pre-Columbus Culture 43
      Indigenous People of the Willamette Valley 44
      Indigenous People of the Spokane Valley 45
      Summary 45
    Colonization 46
      First Encounters with Whites and the Establishment of Trading 46
      Early Missionaries 47
      Early Laws and Congressional Acts 47
      Spokane Reservation 49
      Later Congressional Acts 50
      Summary 51
    Land Use Beliefs and Industrial Interests 51
      Historical Preservation in Millersburg and Albany 52
      Historical Preservation in Wellpinit and Ford 52
      Acceptable Industries in Millersburg and Albany 54
      Acceptable Industries in Wellpinit and Ford 56
      Competing Interests in Millersburg and Albany 57
      Competing Interests in Wellpinit and Ford 59
      Decision-making Time Frames in Millersburg and Albany 60
      Decision-making Time Frames in Wellpinit and Ford 61
      Trust in Millersburg and Albany Decision-makers 61
      Trust in Wellpinit and Ford Decision-makers 61
      Summary 62
    Preliminary Conclusions 64
    Notes 65

5   Defining Environmental Risks:
The Case of Teledyne Wah Chang Albany, Oregon
    Introduction 67
    Teledyne Wah Chang Albany (TWCA) Production Origins 67
      Physical Characteristics of TWCA Operations 69
      TWCA's Environmental Legacy 69
      TWCA Superfund Waste Issues 70
    Hazard Ranking Scores for TWCA 71
    Interpretations of Risks Associated with TWCA 72
    Mitigation Actions for TWCA 79
    Preliminary Conclusions 81
    Notes 84

6   Defining Environmental Risks:
     The Case of the Dawn Mining Company, Washington
         Introduction   85
         Dawn Mining Company (DMC) Production Origins   86
             Physical Characteristics of DMC Operations   88
             DMC's Environmental Legacy   89
             DMC Superfund Waste Issues   90
             DMC Non-superfund Waste Issues at the Mill Site   91
         Hazard Ranking Scores for DMC   93
         Interpretations of Risks Associated with DMC's Midnite Mine   95
         Interpretations of Risks Associated with DMC's Mill Site   97
         Mitigation Goals for DMC's Midnite Mine   100
         Mitigation Goals for DMC's Mill Site   101
         Preliminary Conclusions   105

7   Public Participation in Environmental Decision-making:
     What is it and When Does it Work?
         Introduction   110
         Lines of Authority in Millersburg and Albany   111
         Lines of Authority in Wellpinit and Ford   112
         Frequency of Participation in Millersburg and Albany   114
         Frequency of Participation in Wellpinit and Ford   114
         Factors that Decrease Participation in Millersburg and Albany   116
         Factors that Decrease Participation in Wellpinit and Ford   117
         Factors that Increase Participation in Millersburg and Albany   119
         Factors that Increase Participation in Wellpinit and Ford   120
         Benefits of Participation in Millersburg and Albany   121
         Benefits of Participation in Wellpinit and Ford   121
         Ways to Improve Participation in Millersburg and Albany   122
         Ways to Improve Participation in Wellpinit and Ford   123
         Preliminary Conclusions   123

8   Lessons Learned about Community Structure and Environmental
     Decision-making: Where do we go from here?
         Introduction   126
         Shared History and Community Identity   126
         Control in Local Decisions and Distribution of Local Power   130
         Participation in Decisions   132
         An Evaluation of the Operational Premises   133
             Operational Premise 1   133
             Operational Premise 2   134
             Operational Premise 3   135
         Where Do We Go From Here?   136
             Recommendations   138
             Final Thoughts   138

Appendices
    A. Key Informant Questions   141
    B. Detailed Discussion of the Colonization of the Oregon Territory   142
    C. Hazards Associated with Teledyne Wah Chang Albany Operations   158
    D. Hazards Associated with Dawn Mining Company Operations   165
    E. Maps and Photos of Teledyne Wah Chang Albany Operations
       and Area   181
    F. Maps and Photos of Dawn Mining Company Operations and Area   184

Bibliography   *190*
Index   *204*

List of Tables

Table 2.1	Distribution of NPL Sites in Oregon and Washington by Status	26
Table C.1	Teledyne Wah Chang Albany (TWCA) Hazard Ranking System Scores	158
Table C.2	Organic substances in groundwater at the TWCA Superfund Site	159
Table C.3	Inorganic substances in groundwater at the TWCA Superfund Site	160
Table C.4	Substances in surface water at the TWCA Superfund Site	161
Table C.5	Substances in main plant containment ponds at the TWCA Superfund Site	162
Table C.6	Substances in surface soils at the TWCA Superfund Site	163
Table C.7	Substances in subsurface soils at the TWCA Superfund Site	164
Table C.8	Half-life and Emitter Types of Radioactive Elements Present	164
Table D.1	Annual Precipitation at the Midnite Mine and Surrounding Area	165
Table D.2	Distance between Midnite Mine, DMC Mill site and Spokane, Washington	166
Table D.3	Elevation of the Midnite Mine Site	167
Table D.4	Description of the Midnite Mine Ore Body	168
Table D.5	Substances present in Pit 3 water at the Midnite Mine	169
Table D.6	Substances present in Pit 3 sediments at the Midnite Mine	170
Table D.7	Substances present in Pit 4 water at the Midnite Mine	171
Table D.8	Substances present in Pit 4 sediments at the Midnite Mine	171
Table D.9	Substances present in Pollution Control Pond water at the Midnite Mine	172
Table D.10	Substances present in Pollution Control Pond sediments at the Midnite Mine	173
Table D.11	Substances present in Blood Pool water at the Midnite Mine	174
Table D.12	Substances present in Blood Pool sediments at the Midnite Mine	175
Table D.13	Substances present in drainage waters at the Midnite Mine	176
Table D.14	Substances present in drainage sediments at the Midnite Mine	177
Table D.15	Substances present in other waterways at the Midnite Mine	179
Table D.15	Other substances of concern present at the DMC mill site	180

List of Figures

Figure E.1   Map of Teledyne Wah Chang Superfund Site and surrounding   181
              area
Figure E.2   Map of Teledyne Wah Chang Superfund Site and surrounding   182
              area
Figure E.3   Aerial view of Millersburg, Oregon   183
Figure F.1   Map of Dawn Mining Company's Midnite Mine, mill site and   184
              surrounding area
Figure F.2   Map of Midnite Mine drainages   185
Figure F.3   Map of Dawn Mining Company's mill site   186
Figure F.4   The town of Ford, Washington   187
Figure F.5   The old school, Ford, Washington   187
Figure F.6   The area near the Dawn Mining Company mill site,   188
              Washington
Figure F.7   Midnite Mine, Pit 3   188
Figure F.8   Midnite Mine, Pit 3   189
Figure F.9   Midnite Mine entrance   189

Preface

Today's increasingly complex relationships between humans and their physical environments introduce new challenges to communities, many of which are already struggling to balance competing demands and environmental risks with limited resources. But what social factors frame community-level decision-making about environmental risks? What is the most effective approach to identify environmental risks and formulate community-level responses to them? These are the questions this book explores by examining how environmental regulators, industrial entities and community members work together to define community and community impacts associated with long-term environmental hazards. Through the review of historical documents and key informant interview data, the specific communities studied are: 1) Millersburg and Albany, Oregon in association with the Teledyne Wah Chang Superfund Site, a metal alloy producer; and 2) Wellpinit and Ford, Washington in association with the Dawn Mining Company Midnite Mine Superfund Site, a former uranium extraction operation on the Spokane Indian Reservation, and its corresponding mill site in Ford.

In part, formulating responses to environmental risks involves uncovering existing and new knowledge about potential harms to our health and ecological environs. As the *social amplification of risk* framework proposes, this can take place through a variety of mechanisms including personal experiences, experiences shared by family, friends and others, and participating in risk decision-making processes. However, we experience risks not only as individuals but also as members of groups. This requires us to think beyond more traditional research approaches that focus on individual responses and consider how social structures shape group behavior. We must also recognize that when our understanding of how the wastes in question impact human and ecological health is not complete, as is the case here, defining affected parties and effects becomes particularly difficult and subject to lengthy debates. Moreover, if the structure of decision-making processes about locally based risks creates divisions among community members, community-level responses to risks may be seemingly impossible to formulate. In spite of such circumstances, and regardless of how limited our knowledge or resources to mange these risks may be, their perpetual occupation in our landscape requires our surveillance indefinitely. This makes understanding community-level decision-making processes all the more important.

With that in mind, chapter 1 reviews community features that the social constructionism, disaster research, risk, and community theory literatures suggest play a significant role in community-level decision-making. Broader features of community life not typically utilized by environmental policies and discussed in this chapter include: 1) shared history; 2) community identity; 3) control in local decisions; 4) distribution of power among local institutions; and 5) participation in decisions about environmental risks and mitigation.

The purpose of chapter 2 is to describe the methods utilized in this project including case selection, operational premises, data collection and data analysis. Considering that knowledge about community-level decision-making processes is minimal, this project employs a strongly qualitative strategy to allow for theory generation. Historical documents and key informant interview techniques utilized to meet these tasks are discussed in detail.

Chapter 3 provides an overview of the processes that decision-makers utilize to define population impacts associated with long-term hazardous wastes in the context of the Superfund Program. By exploring how the Superfund Program scores, ranks and selects sites that merit federal intervention, this chapter sheds light on how the structure of decision-making processes may enable or constrain of community-level responses. This overview also prepares the reader for the two case studies that follow.

Chapter 4 assesses how the shared history of community members in Albany and Millersburg, Oregon, and Ford and Wellipint, Washington shaped community identities. Historical factors that significantly contributed to the formation of community identities include geography, the Pre-Columbus culture, colonization, land use beliefs and industrial interests. As the data illustrate, understanding a community's shared history and identity will help decision-makers be aware of potential conflicts that may impede decision-making processes.

Chapter 5 describes how the special metals alloy production facility, Teledyne Wah Chang Albany, became a part of the culture and landscape of Millersburg and Albany, Oregon. It also analyzes how community members, industrial entities and decision-makers define population impacts associated with wastes generated by Teledyne Wah Chang Albany operations between 1956 and 1980. Specific items discussed include the production origins of Teledyne Wah Chang Albany, the identification of waste issues and hazard raking scores for Teledyne Wah Chang Albany that led to its placement on the National Priorities List, interpretations of ecological and human health risks, and ideas about mitigation strategies. Our understanding of how these wastes impact human and ecological health, both as individual components but especially in combination, is not complete. Such circumstances can generate multiple interpretations and differences in opinions about mitigation needs. By evaluating these different ideas, this chapter will provide insights about environmental decision-making processes and identify opportunities for improvement.

Chapter 6 traces how uranium mining became a part of the culture and landscape of the Spokane Indian Reservation and the near by community of Ford in the state of Washington. It also investigates how community members, industrial entities and decision-makers define population impacts associated with wastes generated by Dawn Mining Company operations between 1954 and 1981. Specific items discussed include the production origins of the Dawn Mining Company, the identification of waste issues and hazard raking scores for the Midnite Mine that led to its placement on the National Priorities List, management of the Ford mill site at the state level, interpretations of ecological as well as human health risks, and ideas about mitigation strategies. In addition, this chapter explores responses to a second former uranium operation on the Spokane Indian Reservation run by Western Nuclear, the Sherwood Mine and Mill so named after the tribe's chief, that is now the only successful mining reclamation project in the Pacific Northwest. As the data illustrate, our poor and incomplete knowledge about potential impacts from these wastes can inspire contentious and lengthy debates. Increasing our understanding of the decision-making processes in this case, however, may provide insights about how to improve risk management strategies elsewhere.

Chapter 7 evaluates how community members, environmental regulators and industrial representatives in Albany and Millersburg, Oregon, and Ford and Wellpinit, Washington participated in decision-making about the Teledyne Wah Chang Albany Superfund site and Dawn Mining Company's Midnite Mine Superfund Site and mill site, respectively. Factors of particular interest include lines of authority, frequency of participation, factors that decrease participation, factors that increase participation, benefits from participating, and ways to improve participation. These insights suggest that understanding how participants interact helps identify potential conflicts that may impede decision-making and provides opportunities to improve risk communication strategies.

Chapter 8 summarizes the roles of the five community characteristics studied and their interactions with respect to the operational premises presented in chapter two. Lastly, this chapter provides recommendations for improving community involvement in community-level decision-making about environmental risks. Increasing our understanding of these important social relationships may better equip both decision-makers and communities to effectively manage environmental risks. This is especially important for the communities in this project, as the types of risks targeted will be present in their landscapes indefinitely.

Acknowledgements

This book is but one product of a life-long pursuit for knowledge, guided and encouraged by many. First, I would like to recognize the scientific and theoretical contributions made by those associated with all the works cited. The intersections of these literatures are the very foundation that made this project possible. Persons that have contributed editorial comments and advice to this project include Marilyn Aronoff, Pam Bertram, Gail Beyers, Michael Brown, Lewis Carter, James Flynn, Lee Freese, Eldon Franz, Barry Green, Barbara Harper, William Hendrix, Paul Hirt, Annabel Kirschner, Fredrick Kirschner, Kazumi Kondoh, Steven Kroll-Smith, Loren Lutzenhiser, Donald MacGregor, Marta Maldonado, Aaron McCright, Christine Oakley, James Rice, Julie Rice, Eugene Rosa, Thomas Rotolo, Terre Satterfield, Paul Slovic, Chad Smith, Orlan Svingen, Jean Wardwell, the late John Wardwell, Edward Weber, David Wherry, Shannon Work and anonymous reviewers. This project is richer for your insights, thank you. I would also like to thank Audrey Crow and her computer support staff at West Shore Community College for their assistance with preparing the final print copy of this text. Finally, I would like to thank the Washington State University Thomas S. Foley Institute for Public Policy and Public Service for their support through the Scott and Betty Lukins Graduate Scholarship and the Washington State University Department of Rural Sociology for their support through the Alexander A. Smick Scholarship for Community Service and Development.

List of Abbreviations

AEC	Atomic Energy Commission
BIA	Bureau of Indian Affairs
BLM	Bureau of Land Management
CERCLA	Comprehensive Environmental Response, Compensation and Liability Act of 1980
CERCLIS	Comprehensive Environmental Response, Compensation and Liability Information System
CFR	Code of Federal Regulations
DDE	Dichlorodiphenyldichloroethylene
DDT	Dichlorodiphenyltrichloroethane
DMC	Dawn Mining Company
EIS	Environmental Impact Statement
EPA	Environmental Protection Agency
ESI	Expanded Site Inspection Report
HRS	Hazard Ranking System
LCMC	Local Citizens Monitoring Committee for DMC mill site, Ford, Washington
LRSP	Lower River Solids Pond at the TWCA Superfund Site, Millersburg, Oregon
NCP	National Contingency Plan
NEPA	National Environmental Policy Act of 1969
NPDES	National Pollutant Discharge Elimination System

NPL	National Priorities List
ODEQ	Oregon Department of Environmental Quality
PCB	Polychlorinatedbiphenyl
RCRA	Resource Conservation and Recovery Act of 1976
RME	Reasonable Maximum Exposure
ROD	Record of Decision
START	Superfund Technical Assistance Response Team
TAG	Technical Assistance Grants
TDA	Tailings Disposal Area at the DMC mill site, Ford, Washington
TWCA	Teledyne Wah Chang Albany
WDOH	Washington Department of Health
WDSHS	Washington Department of Social and Health Services

$\mu g/l$ = micrograms per liter

mg/kg = milligrams per kilograms

pCi/l = picocuries per liter

pCi/kg = picocuries per kilogram

Chapter 1

Bringing Community into Environmental Decision-making: An Overview of the Theoretical Orientation

Can we know the risks we face, now or in the future? No, we cannot; but yes, we must act as if we do. Some dangers are unknown; others are known, but not by us because no one person can know everything. Most people cannot be aware of most dangers at most times. Hence, no one can calculate precisely the total risk to be faced. How then, do people decide which risks to take and which to ignore? On what basis are certain dangers guarded against and others relegated to secondary status? (Douglas and Wildavsky, 1983, p. 1).

Introduction

As the intimate connections between humans and their physical environments grow increasingly complex in the present 'era of mega risks' and technological obsessions (Beck, 1992), communities struggle to balance economic interests, development desires and survival needs. These complicated decision-making processes may encompass risks knowingly accepted as well as unwanted risks resulting from the actions of powerful private or industrial interests, including risks imposed by outside entities. Bryant (1995) goes so far as to suggest that some communities may be deliberately targeted for specific types of development and even undesirable land use. For example, both poor and minority communities bear a disproportionately high burden of land fills and hazardous sites. Reasons given for such inequitable land use include the availability of cheap land, the wish to concentrate contaminants in areas already polluted and out of the backyards of those more powerful, industrial encroachment coupled with decreased mobility among the poor and minority groups, the attraction of affordable housing for low-income residents accompanied with the absence of hazard disclosures, and the exclusion of the poor and minority groups from environmental decision-making processes (Bullard, 1990; Been, 1994; Mohia ,1995; Pellow, 2002; Pellow and Park, 2002). None of these issues are easy to resolve nor does the complexity of environmental decision-making end there. For example, new knowledge about human-environment relationships defines and redefines risks, questioning the acceptability of past practices. In other situations, critical information about hazards does not exist, leaving individuals and communities uncertain about what the future may hold.

As we face these uncertainties and challenges, one thing is clear: the process of living in this complex society requires that we prioritize risks and make decisions about how to respond to them not only as individuals but also as communities. Time constraints and a lack of resources may, however, inhibit the extent to which community members can participate in and influence decision-making processes, requiring them to rely on technical experts and environmental regulators in new ways (Chess et al., 1995; McGee, 1999). Moreover, the focus on 'community' as the unit of impact and recent efforts to improve community involvement in many environmental policies calls for an understanding of community features that frame and are affected by environmental risk decisions. This is especially important when differences in how environmental regulators, industrial entities and community members frame risks result in significant conflicts that delay necessary mitigation actions (Gray, 2003; Putman and Wondolleck, 2003). But what do we know about community-level decision-making strategies in regards to environmental risks? What community attributes should we measure to most effectively identify community impacts and yield the most reasonably informed verdicts? Where do we begin?

In epidemiology, a common starting point for identifying the particulars of a suspected risk is to describe its person, place and time characteristics in efforts to identify biologically plausible disease mechanisms (Wing, 1994). Strictly focusing on biological processes, however, creates a tendency to neglect important social features that help explain how the presence of risks may or may not produce undesirable outcomes. This fact, coupled with gaps in biological knowledge, tends to place natural sciences on a collision course with the social sciences. Along the way, we find that sociology offers some important insights about how environmental regulators, industrial entities and communities establish risk priorities and formulate responses to risks in the subspecialties of social constructionism, disaster research, risk, and community theory. Unfortunately each of these literatures is contained within its own niche and pursues its own interests, failing to advance a comprehensive theory that specifically targets the questions above. In order to begin answering the pivotal questions that lay before us, this chapter will integrate important community features in a way that will allow us to better understand community-level responses to environmental risks.

Before proceeding, however, it is important to recognize that properly determining how environmental risks impact communities is not a new or isolated challenge. The National Environmental Protection Act (NEPA) signed by President Nixon on January, 1, 1970, was the first of many national policies to formally require an evaluation of how proposed actions may impact local environments[1] (Fogleman, 1990). In 1980, the Comprehensive Environmental Response, Compensation, and Liability Act (CERCLA), more commonly known as Superfund and administered by the Environmental Protection Agency (EPA), became the first congressional act to address abandoned and uncontrolled hazardous wastes (EPA, 2000a, 2003a).

Although the National Priorities List (NPL) developed by the Superfund program names over 1,500 sites under consideration for federal mitigation, this represents less than five percent[2] of all hazardous sites identified (EPA, 2003a) and suggests that managing environmental risks on some level is an increasingly commonplace task for communities. Hazards of greatest concern under Superfund involve the deposition of pesticides, heavy metals and radionuclides in air, soil and food sources, potentially affecting large numbers of people and ecological systems indeterminately (Finkel and Golding, 1994). Thus, effective mitigation planning requires the consideration of potential impacts to communities that lie in the exposure pathways of suspected risks. This implies that the thorough identification of such impacts largely rests upon how we define 'community' and makes the task of conceptualizing 'community' important for two primary reasons.

First, the potential impacts of environmental risks extend to not only individuals and ecological systems, but also infrastructures and collective social environments —including communities. Admittedly, we do not fully understand all of these impacts much less how the impacts interact. We do know, however, that a number of federal environmental regulations and policies mandate public involvement when evaluating potential impacts of proposed actions and uncontrolled wastes. For example, NEPA specifically requires the consideration of how proposed actions may disrupt community cohesion and desirable growth (Cleckley, 1997). Moreover, the EPA's policies for public participation and consultation with tribes and indigenous groups direct decision-makers to know the affected parties, involve affected parties as soon as possible, and establish ongoing relationships built on honesty and integrity with them (EPA, 2000b; National Environmental Justice Advisory Council, 2000). While these policies offer sound advice, the extent and manner in which federal agencies carry out these activities is up to their discretion. This leaves ample room to formulate many different ideas about what constitutes a community. If such processes result in conceptions of 'community' that are too restrictive, significant impacts and affected areas may be overlooked, rendering mitigation plans unsuccessful. Therefore, in order to avoid making such mistakes, albeit unintentional, we must be mindful of how we define 'community'.

Secondly, community involvement plays an important role in mitigation from both an assessment and intervention perspective. With respect to assessment, community members may provide invaluable information about important impacts that decision-makers unfamiliar with the affected location may otherwise exclude or improperly estimate (Carr and Halvoren, 2001). Hence, local knowledge of previous and current environmental and social conditions is necessary to properly identify the scope of potential impacts. Such knowledge also helps us understand how the acceptability of past practices changes over time and may reframe old issues as new problems. In order to fully utilize the knowledge available, however, one must consider what community entities are included and excluded from assessment and intervention planning. If affected groups do not have the opportunity or resources to participate in risk characterization activities, conclusions about potential impacts may not only be incomplete but may also significantly compromise the success of planned interventions and delay or prevent the realization of long-term mitigation benefits. Thus, in order to carry out policies effectively regulatory agencies need to pay close attention to how they define and involve the affected 'community'.

There are some problems with current environmental risk management and research approaches that limit their ability to make recommendations about how to improve community involvement strategies, however. For example, risk research mostly focuses on individual risk perceptions, individual decision-making, expert opinions and compliance to the exclusion of communities. This creates an opportunity to underestimate the role of group dynamics in decision-making processes and yields little specific knowledge about community-level responses to environmental risks. Furthermore, environmental regulators focus on the physical parameters of the contaminants present. For example, Superfund largely limits the definition of community to persons residing within ascribed physical boundaries in which ecological risks are present (Center for Hazardous Waste Management, 1989). While this is a reasonable place to begin ecological characterizations, it is a very narrow definition of community. Considering only the impacts to such a narrowly defined geographical community leaves us wondering if we have correctly identified both the population affected by environmental risks and mitigation decisions, and the magnitude of the impacts. Such incomplete information may also impede our understanding of community impacts and in turn, limit mitigation success. This suggests that the inadequacies of current research and assessment strategies require a new approach for examining community-level responses to environmental risks. In order to address these shortcomings and better detect community impacts, any new approach must encompass broader features of community if it is to improve risk management techniques.

We can begin developing such a comprehensive approach by identifying community characteristics that provide a structured framework for community-level decision-making about environmental risks. To that end, sociological theory offers some insights about what community attributes may frame and be affected by environmental risk decision-making processes. More specifically, community attributes discussed in this chapter and suggested to be of particular importance by the social constructionism, disaster research, risk and community theory literatures include: 1) shared history; 2) community identity (e.g., geographical boundaries, historical images, physical structures, stigma effects, and attachment to place); 3) control in local decisions; 4) distribution of power among local institutions; and 5) participation in decision-making activities. Developing a better understanding of how communities define and organize themselves, and how environmental risks and mitigation efforts may disrupt community cohesion, i.e., the ability of community members to sustain a common life and form collective responses, will allow us to engage in substantive mitigation policy.

Shared History

Traditionally, community theorists examine how a specific location with a common set of characteristics shapes human behavior. Generally speaking, this strategy identifies a specific geographical location as a community and then ascribes unique characteristics to it, distinguishing rural and urban settings. Characterizations reserved for rural communities include natural resource and agriculturally-based settlements where members share strong ties out of kinship and necessity, practice similar traditions within clearly defined roles, have a strong attachment to place and tend to view outsiders with suspicion. Urban communities on the other hand, are places of high mobility and heterogeneity, and home to multiple impersonal bounds (Bell and Newby, 1971; Carroll, 1995; Lynd and Lynd, 1929, 1937; Gans, 1962; Peterson, 1987; Redfield, 1947). The advantage of defining community in terms of place-based characteristics is that it provides the opportunity to study how physical environments shape and organize human behavior, further enabling correlations between specific social types and bio-physical conditions (Park et al., 1925). Said another way, as we better understand how humans interact with and attach meaning to their physical environs, we can better identify how and what social structures arise that constrain and enable human behavior.

These place-based characterizations also suggest people within fixed geographic locations share similar experiences over time, making shared history potentially an important community attribute. Building upon this idea, other community theorists propose specific attributes for community. These approaches posit the existence of an ideal type of community, capable of sustaining a common life through its richly cohesive social and biophysical arrangements. A community's history (e.g., historical boundaries, demography and geography in conjunction with customs, language, institutional life, and a legacy of events and crises) for example, may provide a sense of rootedness, belonging and commitment among community members (Selznick, 1992). In this sense, history constructs a dialogue that both establishes and reproduces beliefs (Cottrell, 1977). Place-based contexts may help reinforce meanings associated with a community attribute, e.g., landmarks depicting shared history. The significance of a community attribute may, however, remain salient even when specific meanings associated with the attribute vary across contexts or from community to community. Identifying community attributes, like shared history, that remain salient across contexts will provide particularly useful guidelines about what social features frame community-level decision-making processes in a universal way.

Exploring this notion somewhat further, human ecologists propose that as people interact with the physical environments in which they live, they create coexisting social processes that interpret and attach meanings to ecological processes (Freese, 1997). The dialogue that develops among people as they interact with common physical environments may then serve as a mechanism to construct ideas about what events are, and are not, historically significant (Slezak, 1994). Such a dialogue also provides an opportunity for community members to engage in claims making activities where they may construct some events as disasters and specific risks as harmful

(Hannigan, 1995). When community members share many collective experiences and engage in a common dialogue that expresses uniform commitment to ideas and cultural beliefs, history serves as the foundation for community cohesion and collective responses (Brown et al., 1989; Cottrell, 1977; Etizioni, 1993, 1996; Kemmis, 1990, 1995; May, 1994; Goodsell, 2000; Selznick, 1992). This includes collective responses to risks. Residence and employment in a common location with clear and distinctive boundaries, coupled with a strong attachment to place, may further reinforce community cohesion, and in turn, enable community members to act collectively (Carroll, 1995; Etizioni, 1993; Redfield, 1947; Selznic, 1992).

As the interdependence between rural and urban communities increases in response to development expansions, advancements in transportation, and the utilization of new technologies, rural-urban characterizations, however, become less clear. Under such circumstances, treating community as only a geographical analytical unit may overlook important community transformations. Such transformations may assign both rural and urban features to specific locations, blurring place-based distinctions and changing how people think about their community (Stein, 1960; Stern, 1993; Allen and Dillman, 1994). In turn, new opportunities for social interaction may change the types of experiences and dialogues community members' share. For example, when people become members of multiple communities, one's history may involve people from other places rather than only shared with place-based neighbors (Wood and Judikis, 2002). At the same time, changes in socially relevant functions of geographic locales may redefine traditional community goals and boundaries (Imbroscio, 1997; Warren, 1977; Wilkinson, 1991).

Furthermore, the historical role of environmental risks may shape community responses in different ways. For example, community members may uniformly perceive environmental risks associated with long-standing industrial operations that historically provided economic benefits to the community as acceptable (Freundenburg and Pastor, 1992; Sokolowska and Tyszka, 1995). This may encourage resistance to mitigation, especially when mitigation or other interventions infer community-level job losses and development restrictions. When environmental risks equally harm community members and traditional ways of life though, mitigation may be a vehicle to restore community cohesion. In this case, community members may readily support proposed interventions that restore social order and traditions (Axelrod, 1994).

Uncertain or incomplete knowledge about human-environment interactions on the other hand, may instigate debates about the source and significance of environmental risks. Through such debates, we draw upon a mixture of scientific information, local knowledge, uncertainty and socially constructed meanings to form perceptions about environmental risks (Anderson, 1997; Hannigan, 1995; Rosa, 1998). We engage in these short-term and long-term interpretive processes not only as individuals, but also as members of groups and communities. That is not to say only those risks that we are aware of and define as problematic are cause for harm. Many environmental risks produced by industrial activities and other human-environmental interactions impact both human and ecological health even if we do not have the means to see, smell, taste or quantitatively identify them (Beck, 1992, 1995; Carson, 1962; Erikson, 1994; Steingraber, 1998). Since we cannot fully detect all risks and their specific outcomes, critiquing scientific information is a worthy endeavor as such processes may produce new knowledge about how the world works (Jaeger et al., 2001; Wynne, 1992). We cannot, however, lose sight of the fact that both the objective reality of environmental risks and social structures constrain interpretations of human-environment interactions (Dietz et al., 1989). In fact, interpretations of human-environment interactions may serve as historical accounts that structure how we assess and define current events. For this reason, understanding the historical framework that contributes to decision-making will shed light on risk debates.

In efforts to understand how community members may interpret historical events as hazardous, we can draw upon the insights of disaster research. Disaster researchers examine responses to location and time-specific events that cause serious disruptions in social order. Disruptions in safety, protection, and the everyday process of living resulting from these events are so significant that they call for internal and/or external intervention (Quartanelli, 1998). In short, these events remind us of how much we take infrastructures and daily social interactions for granted. Hence, disaster research provides guidance about what impacts community members may deem most historically significant and the degree to which community members experience risks collectively. For example, factors associated with increased psychological severity of disasters include property loss, less distance between one's residence and the site of the event, longer length of residence, separation from family members, presence of young children, lower socioeconomic status, permanent relocation, injury and death (Everest, 1986; Fowlkes and Miller, 1982; Levine and Stone, 1986; Erikson, 1994; Dalton et al., 1999; Brown et al., 2000). As this suggests, the more uniformly traumatic and catastrophic the event, the more likely it is to be incorporated into a community's dialogue and shared risk history.

One must not overlook the important role that physical characteristics of landscapes play in defining environmental risks and developing mitigation options as they influence the types of community transformations that are possible at least to some degree (Cairns and Niederlehner, 1996; Forman, 1995). If, for example, environmental risks impact social and biophysical features in ways that encourage short residential durations, it may be difficult to maintain family ties, group activities, customs, language, and institutional life (Yen and Syme, 1999). Put another way, social disruption resulting from environmental risks may make historical connections increasingly difficult to preserve, leaving the community divided in unexpected ways. Under such circumstances, collective decisions about environmental risks may be seemingly impossible to make (McGee, 1999). On the other hand, when community members uniformly share similar experiences over extended periods of time, cohesive relationships between members may promote quick and consensual community-level decision-making. This makes the type and degree to which community members share experiences fundamental components of the community-level decision-making framework. It also makes understanding differences in dialogues critical for identifying sources of conflict and agreement in decision-making processes. Thus, recognizing the role of shared history in decision-making is essential for properly identifying community impacts resulting from environmental risks and formulating appropriate risk management strategies in a timely fashion.

Community Identity

Community theorists propose that symbols and physical structures help produce and reproduce community social boundaries and the identities associated with them (Cohen, 1985; Freie, 1998; Petersen, 1987). More specifically, community identity emerges from interactions among community members, the integration of demographic composition, organizational networks, industrial and economic components, authoritative structures, functions, norms, common values, symbols, objects, buildings and geographic features (Cohen, 1985; Filkins at al., 2000; Kaufman, 1977; Lynch, 1988; Selznick, 1992). In this sense, community identity is the product of past and ongoing negotiations, and formulating responses to risks is part of that process. That being the case, striking a balance between a unique community character and diverse, integrative attitudes in order to sustain a common life poses challenging and potentially contentious issues. In fact, disagreement about meanings, changes in the functions of physical structures, or the removal of key physical structures and landmarks, may fracture a community's identity (Jenkins, 1996; Rubin, 1977). At the same time, rigid community identities may be difficult to maintain in light of changing circumstances and indirectly erode community cohesion (Goodsell, 2000; Selznick, 1992).

In order to understand how a community constructs its identity, one must unravel the meanings community members attach to physical and social structures. Such meanings may influence how community members interpret environmental risks. To that end, disaster researchers provide insight about how communities formulate conflicting risk interpretations. When the harmfulness of a set of environmental conditions is vague, claims making activities about the significance of risks and how to respond to them compete with one another (Hannigan, 1995). Moreover, different ideas about community identity may incorporate risks in a variety of ways, further contributing to such debates. For example, contaminants associated with economic prosperity may be insignificant in the minds of those that benefited from their creation. These same contaminants and the landmarks associated with them may serve as reminders of personal illness and loss of culture for others. Thus, that smell of money for some may be the smell of disease, death and a host of personal losses for others.

Incomplete information tends to preclude resolution of which claim and/or identity construction is most correct (Couch and Kroll-Smith, 1994). For example, in order to classify a site for inclusion on the National Priorities List, a threat to environmental and public health must be evident and the magnitude of the threat must be severe enough to warrant federal intervention (Landy et al., 1994). The consequences of some contaminants may be uncertain, however, making it difficult to precisely determine who is at risk and the severity of potential impacts. Other contaminants may be undetectable altogether. This creates a situation where multiple interpretations of available physical evidence are possible, leaving the door open for a range of conclusions. It also draws attention to the fact that our values, beliefs, cultural experiences, and social interactions shape how we define and respond to risks (Slezak, 1994; Stern and Dietz, 1994). Furthermore, only those risks deemed socially legitimate capture widespread support and become the object of organized responses targeting their minimization or elimination (Hannigan, 1995). For example, it is not merely the occurrence of a seismic event or presence of an environmental hazard that we respond to, but rather the disruption in social order and fracture in community identity resulting from such events.

The risk perception literature provides insights about what types of risks individual community members may be most fearful of. Personal characteristics associated with heightened perceptions of risk, particularly nuclear wastes, include being female (Boholm, 1998; Dunlap et al., 1993; Flynn et al., 1994; Frey, 1993; Newcomb, 1986; Sjöberg and Drtot-Sjöberg, 1991; Slovic, 1992, 1997; Vaughan, 1995; Whiteman et al., 1995), having children (Bellrose and Pilisuk, 1991; Dunlap et al., 1993; Flynn et al., 1994; Frey, 1993; Slovic, 1992, 1997), lower economic and educational status, (Bellrose and Pilisuk ,1991; Boholm, 1998; Dunlap et al., 1993; Flynn et al., 1994; Slovic, 1992, 1997; Sokolowska and Tyszka, 1995), racial minority status (Boholm, 1998; Slovic, 1997; Vaughan, 1995), residing in highly industrialized areas (Goszczynska et al., 1991), and employment in the industry under scrutiny (Rundmo, 1995; Sjöberg and Drtot-Sjöberg, 1991). Individuals do not perceive or experience risks in a vacuum, however, but do so within a social and collective context. For example, some workers are willing to voluntarily accept extra risks in exchange for higher wages (Bellrose and Pilisuk, 1991; Nelkin and Brown, 1994; Vaughan, 1995; Viscusi, 1983). Thus, we must look beyond individual risk responses and understand how community members define and experience environmental risks collectively.

The *social amplification of risk* framework provides us with a starting point for understanding how people interact with and define risks within the context of their community identity. This framework proposes that: 'risk events interact with psychological, social, and cultural processes in ways that can heighten or attenuate public perceptions of risk and related risk behavior' (Kasperson et al., 1988, pp. 178-9). Initially, values, beliefs and social structures filter signals about risks. Thereafter, cognitive processing at individual, organizational, and community levels decode the filtered signals. During this decoding process, interactions with social and environmental structures through both formal and informal communication networks mold behavioral intentions. Both individuals and groups may carry out these plans (Kasperson et al., 1988). In some situations, lack of community-level agreement about risks may encourage affected individuals to treat risks as a personal responsibility, promoting individual rather than collective responses (McGee, 1999). Intended and unintended consequences of responses to risks, e.g., development of new technologies and security procedures, enduring negative images and loss of trust in institutions, may further amplify or attenuate risk decisions (Axelrod, 1994; Clarke, 1989; Dietz et al., 1996; Kasperson et al., 1988; Krimsky and Golding, 1992; Ostry et al., 1995; Renn et al., 1992; Rosa, 1998; Rosa and Dunlap, 1994; Short, 1984, 1992; Weart, 1988). This implies that debates about risks may in turn, be coupled with debates about community identity as risk responses may reformulate community definitions and priorities. On the other hand, desires to maintain community identity may promote actions to down play the significance of risks.

As suggested above, the availability of information complicates this framework. More specifically, risk analyses provide some empirical observations about risk outcomes, but all of the information needed to make thoroughly accurate decisions is frequently not available (Beck, 1995; Rosa, 1998). Furthermore, even under the best circumstances calculated probabilities are only socially constructed predictions, not actual events, and are subject to varying degrees of error and dispute (Merkhofer, 1987). In the case of Superfund for example, the hazard scores assigned to sites under investigation indicate only potential, not fully realized, human and ecological health risks. These hazard scores, each with their own potential flaws, guide decisions about which sites to place on the National Priorities List (Interorganizational Committee on Guidelines and Principles for Social Impact Assessment, 1997). As new information becomes available, risk definitions and hazard scores may be subject to numerous revisions. Given the inherent uncertainty of probability estimates, substances thought of as safe may become significant risks, and vice versa, in the process. At the very least, ample opportunities to debate risk characterizations and NPL status exists.

In the midst of risk debates, symbols, objects and buildings that reinforce community identities and serve as vehicles to promote some arguments over others, may also create stigma and alienate individuals and groups from the majority (Goffman, 1963). As the disaster and risk literatures suggest, this has important implications for communities in Superfund zones. Superfund targets abandoned sites and uncontrolled releases that pose threats to public health such as abandoned manufacturing facilities, processing plants and landfills, where the people responsible for the contamination cannot be found, cannot perform, or cannot pay for clean up work (EPA, 2000a). Associated vacant buildings convey a breakdown in infrastructures; they also convey a breakdown in social structures. If mitigation efforts take several years to complete, disjointed appearances may amplify lost social order (Kroll-Smith and Couch, 1991; Erikson, 1994). For example, avoidance of high-risk areas may erode critical social structures and disrupt traditional ways of life, weakening community cohesion (Satterfield et al., 2000). Such conditions may leave community members struggling to maintain a community identity of better days while that same identity is cause for harm to their health. At the same time, community members may not agree about potential threats to public health, creating divisions between those most and least affected. This may in turn, fracture an already fragile community identity, making collective responses to risks all the more difficult to formulate.

In other scenarios, persons outside of the community may impose ideas of harmful states whereas local residents interpret their surroundings as safe. To minimize stigma, communities may attempt to divorce themselves from historical images by changing themes and creating new appearances in order to interest outside investors and build new economic bases (Kunreuther and Slovic, 1999; Swearengen, 1996). New identities, however, may be sources of conflict in community-level decision-making processes (Bunker Hill Superfund Task Force, 1994). Here too, lack of uniform agreement about the fearfulness of environmental risks may encourage some community members to form new relationships within cohesive subgroups and give rise to new social movements (Clarke, 1989; Cleckley, 1997; Couch and Kroll-Smith, 1990). These additional community divisions may not only further impair community cohesion, but may also make achieving consensus about mitigation planning increasingly difficult. This requires environmental regulators to be knowledgeable of differences in community identity constructions in order to avoid aggravating underlying conflicts.

The presence of environmental risks alone, however, does not assure stigmatization of the surrounding landscape. If, for instance, the visibility of the suspected risks is not obvious, the risks may generate little or no stigma (Jones et al. 1984, Edelstein 1988). Moreover, when local residents have a long-standing knowledge of the risks prior to their formal discovery, risk management activities may have little or no impact on daily social processes (Beamish, 2002; Zavestoski et al., 2002). Under such circumstances, environmental risks may receive little, if any, attention from community members whereas the newly 'discovered' risks may be of grave concern to outsiders, especially when the potential mobility of the contaminants threatens the health of outlying areas, i.e., downwind or downstream communities (Steingraber, 1998). Here again, reaching consensus between environmental regulators and community members about environmental risks and mitigation actions may be difficult. Furthermore, an inadequate understanding of differences in risk perceptions among all parties involved may unintentionally heighten disagreements, making an awareness of community identity all the more important for mitigation planning.

If distrust in environmental regulators and/or industrial entities is part of the community identity, community members may be suspicious of the technical or 'official' risk presentations and draw more heavily upon personal experiences when making decisions. Not only are residents more fearful of hazardous facilities when they do not trust environmental regulators, but as distrust in regulators increases, community members' sense of control over their locale decreases (Frey, 1993). Under such conditions of ingrained distrust, gridlock, or the inability to reach a decision, is even more likely (Rosa and Clark, 1999; Williams, 2002). On the other hand, if a history of cooperation where community members feel decision-makers made an effort to get to know them exists, decisions may be timely (Komter, 2000; McCool and Guthrie, 2001). As this suggests, better understanding a community's identity will allow us to more readily identify sources of potential conflict that may impede decision-making processes. Furthermore, lack of knowledge about how communities frame risks may in fact, hamper risk communication efforts (Chess et al., 1995; Vaughan, 1995). Thus, in order to capitalize on opportunities to improve our risk management strategies, we must consider the role of community identity in community-level decision-making.

Control in Local Decisions

In today's global society, maintaining unique community identities becomes increasingly difficult as community members attend school, work, shop and live in multiple locations. Not only does it become challenging to maintain shared values, beliefs, dialogues and experiences as already discussed, but community members may find it difficult to maintain control in local decisions as external ties grow stronger (Freie, 1998). Since internal community discontinuities may further amplify differences in risk perceptions (Kasperson et al., 1988), understanding how community functions tie together is an important issue for community theory and community-level decision-making. In response to this need, some community theorists treat community as a system of interrelated networks.

Early on, Parsons (1951) for example, characterized community as a system of goals, environmental adaptations and manipulations, and integrated human interaction. Building upon these ideas, Warren (1978) developed a systems approach to describe socially relevant functions with local origins. He defined vertical patterns as hierarchical lines of authority and structures within systems. Horizontal patterns on the other hand, represent how a system relates to other systems, drawing attention to outside influences that may potentially impact internal systems. While vertical ties are important with respect to localized associations, Warren found the loose coupling between horizontal ties that connected communities to other communities, the state and the nation, as equally important. Through these social networks, Warren argues it is possible for organizations to become communities of their own. Community theorists interested in specific attributes that define community go a step further and propose that plurality, the participation in diverse organizations outside the immediate area and different roles that community members take on within them, is fundamentally a critical component of community life (Cottrell, 1977; Kemmis, 1990, 1995; Selznick, 1992). Together, these theoretical approaches suggest a community's external networks play an important role in determining how much control a community has in local decisions.

When external organizations significantly contribute to local community infrastructures, it may be difficult for the community to maintain local control. In the event of a disaster for example, faulty infrastructures, inadequate housing, lack of insurance, and lack of savings prior to the event can amplify impacts and complicate recovery efforts. Assistance from outside agencies may not only be necessary for recovery, but may inadvertently weaken already fractured local community relationships (Kroll-Smith and Couch, 1991). Since the need for such assistance becomes magnified in times of community disaster, more closely examining response patterns to different types of disasters may shed light on how shifts in power from communities to external organizations impact community cohesion and community-level decision making.

With that in mind, some disaster researchers differentiate disasters on the basis of natural versus technological causative agents, suggesting responses to each follow distinctive patterns. Reactions to natural agents such as floods and earthquakes for example, frequently follow a uniform pattern. Suddenly, and with little or no warning, an established social order and familiar daily routines become chaotic and unpredictable. Following the unanticipated, unavoidable, uncontrollable event, internal and/or external agencies work with disaster victims to reconstitute order and restore daily routines (Couch and Kroll-Smith, 1990). Because natural agents oftentimes swiftly affect thousands and personal harm and property damages are readily visible, there is little debate about necessary interventions and who is in need of assistance. Such events frequently overwhelm the capacity of local organizations (who may be victims themselves), rendering them unable to assist all of those in need. As a result, the affected community may require assistance from external organizations (Dombrowsky, 1998; Erikson, 1994). Since no one appears to be directly at fault, uniform and sympathetic responses to clearly defined victims generally restore at least some order in a timely fashion. Less socially powerful groups, however, may be the last to benefit from agency interventions in spite of the fact that they may experience the most significant losses and threats to public health (Brown et al., 2000; Hewitt, 1998).

Responses to technological agents on the other hand, are far less clear. There may be some confidence in the ability to predict the failure of human-made systems and infrastructures (Sagan, 1993; Perrow, 1984, 1994); however, the severity and magnitude of adverse outcomes associated with suspected agents, as well as the agents themselves, are often ambiguous. All social and economic classes may be equally affected, blurring the distinction between who is and who is not in need of assistance (Beck, 1992). Established community institutions and organizations may also become the target of external criticism in efforts to prevent future disasters. Community members may interpret such criticisms as outsiders attacking community identity, which may in turn, may cause additional disruptions in local social order. Such perceived attacks may also encourage community members to question past practices formerly deemed acceptable, introducing more uncertainty to already confusing times. In this cloud of uncertainty, agencies, experts and individuals frequently engage in lengthy debates about the source of the disaster, identification of responsible parties and what interventions are necessary, and for whom (Couch and Kroll-Smith, 1990). Complicating matters further, impacts resulting from technological disasters, such as the improper disposal of chemical wastes at Love Canal, may extend for years without ever reaching clear conclusions (Kroll-Smith and Couch, 1991). In some situations, it may not be possible to restore order following a technological disaster. Without local order, community members may find it difficult to actively participate in local decisions, increasing their sense of powerlessness (Chess et al., 1995; Fowlkes and Miller, 1982; McGee, 1999).

Natural and technological disasters are not necessarily mutually exclusive, however. For example, the effects of an earthquake (natural event) become disastrous only when human-made systems fail, e.g., buildings collapse, gas pipe lines rupture, water and waste disposal systems are destroyed. Otherwise, the earthquake is merely a seismic event that may even go unnoticed. The failure of technological systems in response to natural events may be unpredictable and have both immediate and unexpected chronic effects, creating multiple periods of crisis indeterminately e.g., mine waste contamination following failure of a dam in Buffalo Creek, West Virginia (Erikson, 1994). Where does the line between natural and technological disasters then fall? Quarantelli (1998) suggests that focusing on the qualities of causative agents is a less important sociological endeavor than carefully examining the responses and outcomes to disaster events. Others argue that since technological systems are clearly not part of natural phenomena, 'failing to distinguish hazard-types in some ways obscures choice and agency' (Clarke and Short, 1993, p. 378). At the same time, we must be careful not to overestimate the range of responses to natural and technical disasters as physical environments and social structures constrain options in both instances. Such constraints may place limits on how much control persons directly affected by a disaster are able to maintain in local decisions.

Local control of environmental risks is a particularly important consideration for communities labeled with Superfund status. Here, outside powers and political forces may compete with local power and vice versa. Outside funds are likely to drive mitigation efforts, especially when potentially responsible parties are negligent and local resources are inadequate. Few, if any, communities have millions of surplus dollars to earmark for environmental clean-up as many Superfund sites dictate, like the Midnite Mine on the Spokane Indian Reservation in this project that is estimated to cost as much as $160 million to remediate (Ichniowski, 2001). In the event that local resources are not adequate, mitigation decisions may be made for the community by outsiders in control of external resources rather than by the community (Clarke and Short, 1993). As a result, community members may feel excluded from mitigation decisions. Furthermore, because they feel there is little they can do and restoration efforts oftentimes proceed at a very slow pace outside of their control, community members may become collectively apathetic (McGee, 1999).

Federal spending in the local area may improve the economic conditions for a community (Mencken, 2000). The risk literature suggests, however, new funds flowing into a community as the result of Superfund status may draw unwanted attention from outsiders to environmental risks (Flynn et al., 1998; Kunreuther and Slovic, 1999). Potential stigma resulting from newfound interests in the community by outsiders may further complicate matters and inhibit future economic growth as well as fragment important social networks. All the while, the desire to correct environmental problems competes with the community's desire to be safe. Maintaining traditional community values, a unique community identity, and financial independence in the midst of all of this may become increasingly difficult, especially when community cohesion diminishes (Satterfield et al., 2000). For these reasons, the control community members have and are able to exercise in community-level decision-making is an important consideration, especially for environmental regulators as their prescribed lead authority role may inadvertently amplify already impaired social relationships.

Distribution of Power among Local Institutions

Another way of conceptualizing community involves identifying critical functions and networks that tie communities together with respect to how organizations within a community interact. More specifically, Warren's scheme of horizontal (i.e., external) and vertical (i.e., internal) ties provides an opportunity to identify internal community components most affected by environmental risks, groups most likely to participate in decision-making processes, and shifts in power among local institutions. Awareness of such issues draws attention to a wider array of community groups and how different groups agree or disagree about environmental risks than typically employed by policy makers. This makes understanding internal community dynamics an important task if we are to improve risk management strategies and community involvement in decision-making.

Community theorists interested in specific attributes that define community go a step further and argue that plurality helps diffuse power widely among a variety of organizations and groups. This broader distribution of power may more readily protect individual rights and values, and may increase opportunities to participate in community level decision-making (Howard, 1995; Moon, 1993). Moreover, examining the distribution of power among local institutions may help identify groups and individuals that do not have an equal opportunity to engage in decision-making (Winner, 1993). This is important when one considers that less socially powerful groups may be the last to benefit from agency interventions in spite of the fact that they may experience the most significant losses and threats to public health (Brown et al., 2000). Vertical ties and lines of hierarchy may also depict how conflict generates distance between groups while simultaneously increasing cohesion within groups (Imbroscio, 1997). Such power shifts introduce autonomy, or the nurturing of individual development, as another necessary consideration (Selznick, 1992). Selznick (1992) suggests that perhaps the combination of autonomy and plurality, or a balance between community and personal commitments, is most critical for sustaining community life and distributing power equitably among local organizations.

As discussed earlier, one way to better understand how events shape organizational responses and how organizations interact with each another is to examine responses to disasters (Clarke and Short, 1993). While disaster researchers focus on external organizational responses, they also provide guidance about how organizations within a community interact. For example, disruptions in daily activities and infrastructure inadequacies resulting from disasters indicate a breakdown in community structure. This fragmentation may cause community organizations to compete with each other in new ways for the limited resources that are available, resulting in shifts of local control across agencies. In some instances, the affected community may need to rely on a new set of leaders with specialized skills. Thus, as recovery efforts take place, local power struggles may encourage different community subgroups to organize themselves along lines of disagreement. In the process, less powerful groups may be silenced while others are promoted (Erikson, 1994; Hewitt, 1998; Kroll-Smith and Couch, 1991). Such power struggles may produce an incomplete understanding of community impacts and the magnitude of such impacts. In turn, these gaps in knowledge may impair risk assessment and mitigation activities.

To better understand the finer points of response patterns, some disaster researchers evaluate responses to different types of events, suggesting that specific events amplify local impacts and responses in distinctive ways. Gilbert (1998) for example, proposes a distinction between wars, events involving social vulnerabilities, and events involving uncertainty. In the event of war, community members organize themselves in reference to a common enemy and react to aggressions targeted at the group. Similarly, vulnerable community members, pushed together by a common external danger, formulate relationships they may not otherwise pursue. When the threat of danger is uncertain, debates about who is responsible, the severity and magnitude of problem, the type of intervention needed, and how to organize responses are likely (Gilbert, 1998). Delays in decision-making stemming from such debates may in turn, produce severe social disruption and irreversible fragmentation of community networks (Kroll-Smith and Couch, 1991). In the wake of such uncertainty, the power distribution across local institutions becomes unclear, making the community's future all the less predictable, and community members begin questioning the community's ability to survive (Couch and Kroll-Smith, 1990).

The long-term social disorder resulting from chronic technological disasters where hazard identification and mitigation takes several years or longer, provoke states of seemingly perpetual chaos and may instigate additional and unrelated conflicts, giving rise to social movements. More specifically, the original problem may evolve into a host of unresolved and even unrelated problems. Disenfranchised community members may form coalitions with each other, forming new special interests groups. These new groups may be at odds with established internal and external institutions (Coleman, 1957). This may place community members in a situation where they are forced to pick sides, subdividing the community in unexpected ways (Couch and Kroll-Smith, 1994; Hannigan, 1995). The resulting subdivisions within the community may be permanent, causing the community to take on new forms. In the process, the control that different local groups have in local decisions may shift dramatically, generating more new groups while eliminating others altogether. The underlying conflicts that result from these power shifts may inhibit community members from forming collective responses that effectively manage environmental risks (Gray, 2003). Hence, understanding which groups have the most and least influence is essential for hazard assessment, mitigation planning and policy formulation.

The risk and social construction literatures provide us with some insights about who may have the most influence in environmental decision-making. For example, scientists and experts advance the framing of risks by attempting 'to show that their existing research work contributes to the solution' (Hannigan, 1995, p. 79). Douglas and Wildavsky (1983) argue that such approaches build confidence in experts, further legitimizing their advice. Inadvertently, this legitimized position provides experts with opportunities to dominate the public policy process and silence less powerful voices.

When contradictions among scientists and experts arise, however, the public's feelings of uncertainty about risk outcomes may increase (Beck, 1992).

The media may also play a significant role in framing risks by providing journalists with the opportunity to 'define and redefine social meanings as part of their everyday working routine' (Hannigan, 1995, p. 59). Furthermore, the media tends to over-report rare, unexpected events like nuclear accidents and plane crashes, contributing to the underestimation of significant but more frequent events, like automobile fatalities. The process of dramatizing events 'tends to divorce issues from their wider social and political context, and rarely incorporates a consideration of multiple explanations' (Anderson, 1997, p. 134). As a result, the media can 'produce a false consciousness that legitimizes the position and interests of those who own and control the media (Anderson 1997, 21). While people acquire information from many other sources in addition to the media such as government agencies, coworkers, advocacy groups and community organizations (Freudenburg and Pastor, 1992), the role of media in amplifying risks continues to be significant (Flynn et al., 1998).

Within this rubric of multiple organizations and groups, community members may struggle with their role in decision-making processes in a variety of ways. If community members feel that they have little control over the risks under debate, they may become more fearful of them (Bellrose and Pilisuk, 1991; Flynn et al., 1994; Kleinhesselink and Rosa, 1991; Machlis and Rosa, 1990; Slovic, 1997). Such anxieties may lead to distrust in experts and policy makers (Beck, 1992). Distrust in policy makers may contribute to gridlock, extending decision-making time frames (Rosa and Clark, 1999; Williams, 2002). Decision delays may inadvertently increase community members' sense of powerlessness in decision-making and increase risk concerns, especially among those with fewer community ties (Whiteman et al., 1995). This makes understanding what social structures constrain and enable community responses, and how power is distributed among local institutions, necessary for effective community-level decision-making about environmental risks as a lack of awareness of such issues may unintentionally build upon underlying conflicts and yield improper conclusions about risks.

Participation in Decisions about Environmental Risks and Mitigation

Another area of interest to community theorists is how social interactions transform community and relationships among community members. For example, Durkheim (1933) made a distinction between mechanical and organic solidarity. Mechanical solidarity is the product of taken-for-granted similarities such as functional, almost automatic yet stifling, relationships resulting from kinship ties, and similar traditions, skills and ideas. Organic solidarity embodies differences resulting from intentionally cultivated relationships, largely driven by labor market exchanges, that may expand opportunities and encourage the development of highly specialized skills. Marx too, conveys the notion of organic solidarity in his use of the term 'organic totality' (Gould, 1978; Kirkpatrick, 1986). Within this concept, Marx defines society as an interdependent organic system of components. Each component possesses different functions, but interactions between components define and redefine social structures as well as the components themselves. Treating the components as community characteristics, community exists 'of itself.' Through interaction, however, participants come to know their shared situation, and community exists 'for itself.' Hence, while Marx's focus is on conflict in society, Durkheim focuses on cohesion. Similarly, Weber (1958, 1978) examined lines of authority, associating traditional authority with kinship ties, and legal-rational authority with contractual, market driven labor exchanges. Marx criticized this approach for taking the underlying determinants of these structures for granted (Kirkpatrick, 1986). Be that as it may, all of these approaches suggest community is not just about geographic location alone, but also about social interaction and the ways in which community members participate in those interactions.

Expanding upon these ideas, Tönnies (1940) investigated social interaction differences in small and large places, and changes in the conceptualization of community over time. He developed a model that characterized two different ways in which people relate to each other, Gemeinschaft (translated as community) and Gesellschaft (translated as society). In Gemeinschaft, kinship and close, intimate relationships structure social interactions. A high degree of group homogeneity results from very ordered roles and identities, and ascribed social status. Travel and relocation of group members is minimal and a sentimental attachment to place persists. Within this rubric, homogeneous social interactions among place-based neighbors structure community and community behavior in predictable ways. As the industrial revolution and urbanization evolved, Tönnies argued that Gesellschaft replaced Gemeinschaft. Gesellschaft consists of impersonal contracts, increasingly complex and multiple identities, chaos, a high degree of mobility and heterogeneous group structures. This suggests that individual differences and diverse social arrangements structure community behaviors in potentially very broad terms, producing a wide range of community forms that are subject to continuous transformation.

For some communities, this may mean that over time, social interaction is a more meaningful descriptor than physical location (Doreian and Stokman, 1997; Rifkin, 1995; Schuler, 1996). Nonetheless, close physical proximity support cohesive relationships among community members (Allen and Dillman, 1994). For example, community members that share a strong attachment to place may be more sensitive to local impacts than short-term residents and/or outsiders. In turn, the desires of place-attached community members to influence local decisions, especially those issues that involve environmental degradation, may encourage active participation in decision-making (Vorkin and Riese, 2001).

Taking this a step further, community theorists interested in specific attributes propose that mutuality, or the degree to which people need each other, may foster community cohesion in additional ways. Community members that are significantly dependent on each other for example, may develop enduring bonds and commitments to each other, if only out of necessity (Etizioni, 1996; Nisbet, 1990; Selznick, 1992). Such enduring bonds, however, may encourage community members to express little interest in outsiders' perspectives (Vorkin and Riese, 2001). While this may not be the best precursor for decision-making processes directed by external environmental regulators, cohesive relationships may foster high levels of participation and enable community members to readily reach consensus about risk priorities and collective responses to risks (Buckner, 1988). Here, the community may also treat those members most affected as victims and provide needed assistance. Similarly, vulnerable community members exposed to a common environmental risk may form cohesive relationships with each other that they might not pursue otherwise (Gilbert, 1998). Thus, mutual dependence encourages community members to be active participants in decision-making and place attachment reinforces such cohesive relationships.

If people do not rely upon each other to any significant degree, however, little need and opportunity to participate in a common life may exist. For example, as the size of a community increases and services become decentralized, competing interests and responsibilities may reduce both the need and opportunity for community members to actively participate in place-based community affairs (Kirkpatrick, 1986; Cuba and Hummon, 1993; Freie, 1998). In this sense, a negligible degree of mutuality may depict active community involvement as undesirable. Disproportionate or low levels of participation on the other hand, may indicate a disinterest in the seriousness of environmental risks among place-bound community members that feel they cannot leave the area (Shriver et al., 2000; Whiteman et al., 1995). Inadequate resources to participate, unclear risk management goals, limited response options and lack of commitment by outside environmental agency staff as well as questionable staff experience in risk communication, may also discourage community members from participating in decision-making processes (Chess et al., 1995; McGee, 1999). Such circumstances may unintentionally isolate the persons most affected and exclude them from community-level decision-making. These exclusions may in turn, fragment the community's identity and social networks, diminish community cohesion, and delay mitigation efforts (Tauxe, 1995).

Complicating matters further, external organizations may be insensitive to the roles they play in contributing to community conflict that inadvertently discourage community participation in decision-making. First, environmental policies designate lead authority to the regulatory agency spearheading a specific environmental issue and this assures a top down management approach to community involvement (EPA, 1994). Secondly, the 1946 Administrative Procedure Act requires federal agencies to hold a public meeting before implementing proposed actions or policy changes (EPA, 1994; Williams, 2002). Additional public involvement actions are completed at the lead agency's discretion and 'there is no requirement that agencies actually respond to citizen input' (MacLennan, 1988, p. 69). Typically, this means that regulatory agencies are heavily invested in risk characterization and mitigation planning before holding public meetings. This sets the stage for the perception that public hearings are primarily about telling the public what the lead agency plans on doing, rather than listening and acquiring information from the public to use in mitigation planning. If community members believe that environmental regulators have already determined the significance of risks as well as interventions, they may feel they have no influence in decision-making processes and therefore, see no reason to participate (Smith and McDonough, 2001). Furthermore, public discussions may be particularly trying when agency personnel enter into public involvement activities with preconceived notions about special interests groups being uncooperative (Halvorsen, 2001).

Other lead agency actions may discourage community involvement as well. For example, when agency representatives fail to answer citizens' questions completely or clearly, community members fear information is being withheld, making such agencies less trustworthy (McCool and Guthrie, 2001). If community members do not trust the lead agency and its key decision-makers, the incentive to constructively participate in decision-making may be marginal at best (Rosa and Clark, 1999). In addition, the release of conflicting information about risks may anger community members and leave them wondering what influence they have (Rowan, 1994). Finally, in the absence of quality community involvement, external agencies may misinterpret meeting participants and/or victims as being representative of the community at large, only to build upon preexisting internal conflicts and perpetuate faulty interpretations of incomplete risk assessment information (McCool and Guthrie, 2001). On the other hand, high quality participation activities like focused conversations with facilitators may help reverse distrust in regulatory agencies and increase support for agency decisions (Halvorsen, 2003). Minimizing conflict may in turn shorten mitigation time lines as well as defray mitigation costs.

This background suggests a high level of community participation in environmental decision-making may indicate community members' desire to influence local risk management strategies. Low levels of participation, however, are more difficult to interpret uniformly. Some community members may have little concern for environmental risks that are of great concern to others. Community members may feel that they have little influence in decision-making processes. Others may be apathetic about long-standing problems that receive little attention. Time constraints and limited resources may also inhibit the extent to which community members are able to participate in decision-making activities, especially when mitigation extends across decades, leaving even the most dedicated participants subject to burn out. Hence, identifying the extent to which communities desire to participant in decision-making, participation barriers, and strategies that most effectively encourage community involvement are important tasks if we are to determine how to improve risk management and risk communication strategies. Assuring opportunities for community members to participate in decision-making about environmental risks will allow for a more thorough identification of impacts and thus, will more likely yield successful interventions. To encourage community involvement, however, regulatory agencies will need to establish clear roles for community members in decision-making processes and specify measurable goals for participation so that community members will feel their time, often limited by competing demands, is well utilized. One must also not lose sight of the fact that industrial representatives and environmental regulators are key players in environmental decision-making processes, and they too, need avenues to share undocumented insights and raise questions without retribution as the waters that lie ahead are largely uncharted.

Final Thoughts

Currently, the impacts of environmental risks and mitigation efforts on community-level social interactions are poorly understood. The *social amplification of risk* framework suggests we form ideas about environmental risks as we interact with the biophysical and social features of the places in which we live, work and recreate. We draw upon these same social and biophysical features to form ideas about community. Hence, the characteristics we associate with our community provide a structured framework to guide community-level decision-making. Given today's many competing demands, time constraints, and frequently limited resources to fully participate in decision-making processes, community members often face difficult choices when setting risk priorities. This makes understanding how communities define and organize themselves cohesively, and how environmental risks and mitigation efforts disrupt community cohesion, important for substantive mitigation policy. More specifically, we need to understand how the characteristics that community members draw upon to define their community, frame ideas about risk priorities and collective responses to risks. Impacts to these characteristics will influence a community's ability to sustain cohesion. Furthermore, the manner in which decision-makers define affected communities may shape not only mitigation efforts, but also the degree to which communities support their decisions. This raises a number of important questions that policy makers need to consider as they engage in risk assessment and risk management activities. How do contaminated communities define and organize themselves? What characteristics are most important for achieving community cohesion? How do environmental risks and mitigation efforts interfere with achieving community cohesion? If we simply assume, rather than empirically identify, those dimensions of community most affected by environmental risks, can mitigation strategies be successful?

As policy makers pursue answers to these questions, the approach to community conceptualization provided here may guide their way. This approach steps beyond strictly geographical notions of community and treats community as a dynamic entity that may be redefined over time by social exchanges through internal and external networks. That is not to suggest that the physical features of a community's landscape are unimportant. If we are to understand anything about communities, we must recognize how human interactions with physical environments shape group responses and how specific features may limit possible responses to environmental risks. We may loose track of our dependence on environmental systems as we ship our wastes out and ship necessities in, but these actions do not negate it.

Although challenging, understanding the complexities of decision-making processes is both a worthy and necessary scientific endeavor given the mandated focus on community impacts and community involvement in environmental policies. But in order for community involvement to be a successful component of decision-making, we must understand how communities organize and define themselves around such risks. In part, this requires us to identify community-level structures that filter and decode signals, which in turn shape collective responses to environmental risks. Such organized responses in combination with physical features may be critical to a community's survival and successful implementation of mitigation efforts. The examination of decision-making processes also provides the potential opportunity to distinguish mitigation successes and failures with an eye towards improving risk communication strategies. Understanding such issues will allow us to improve community involvement in mitigation efforts and in turn, increase the likelihood of mitigation success. As we continue to make new discoveries of uncontrolled hazardous wastes and bring into question past as well as current industrial practices, we will continue to be challenged with the management of these important social and environmental issues for many years to come.

Notes

1   In the United States, legislation addressing environmental risks began most notably with the signing of the National Environmental Protection Act (NEPA) on January 1, 1970 by President Nixon. The purpose of NEPA is:

> To declare a national policy which will encourage productive and enjoyable harmony between man and his environment; to promote efforts which will prevent or eliminate damage to the environment and biosphere and stimulate the health and welfare of man; to enrich the understanding of ecological systems and natural resources important to the Nation; and to establish a Council on Environmental Quality.

Simply put, NEPA requires an evaluation of the potential environmental consequences of proposed federally regulated actions. It also requires the public to be informed about the evaluations (Fogleman 1990). However, only those social impacts that can be directly linked to environmental impacts associated with proposed acts require consideration. The NEPA targets proposed activities, however; it does not address pre-1970 actions or impacts related to abandoned facilities and operations. The same can be said of the Clean Air Act of 1970, the Federal Water Pollution Control of 1972 and the Clean Water Act of 1977. While the Resource and Recovery Act (RCRA) of 1976 addressed improved procedures for managing the transportation, treatment, storage and disposal of hazardous wastes generated from concurrent operations, it too does not address the management of previously generated and disposed wastes (Center for Hazardous Waste Management 1989). Many states developed their own interpretation of these acts and in some instances, extended them. The management of previously generated and disposed wastes is not addressed under these acts unless it is directly related to concurrent or proposed activities. The first legislative act that addresses abandoned and uncontrolled wastes is the Comprehensive Environmental Response, Compensation, and Liability Act (CERCLA) of 1980.

2   There are currently 1,235 sites on the NPL with an additional 59 proposed, 216 deleted, and 19 partially deleted, for a total of 1,529 sites (Environmental Protection Agency 2000c). The Comprehensive Environmental Response, Compensation, and Liability Information System (CERCLIS) database is maintained by the Environmental Protection Agency (EPA). This database contains NPL sites as well as potentially hazardous sites requiring preliminary investigation. Currently, there are 453 CERCLIS sites for EPA Region 10 and 10,784 nationally maintained in this database. If one considers archived files, over 42,000 CERCLIS sites have been identified nationally since the establishment of CERCLA. Among both current and archived CERCLIS sites, less that 5% become NPL sites (EPA 2003a). However, as many as 425,000 potential sites have been estimated (Hird 1993).

Chapter 2

Methods:
New Strategies for Old Problems

The real objective of research should not be to repeat old 'truths', it is to find out about new points that contribute to the scientific and public discourse on social phenomena (Alasuutari 1995, 2).

Introduction

This project aims to: 1) identify and better understand how environmental regulators, industrial entities and community members define communities affected by long-term hazardous wastes; 2) better understand how differences in community constructs influence relationships among decision-makers and community members with respect to ownership of environmental threats and mitigation efforts; and 3) better understand how decision-making processes about environmental threats and related mitigation efforts impact community cohesion. Specific community attributes described at length in chapter one that this project evaluates in order to better understand these relationships are: 1) shared history; 2) community identity; 3) control in local decisions; 4) distribution of power among local institutions; and 5) participation in decisions about environmental risks and mitigation. The purpose of this chapter is to describe the methods and design components utilized to examine these issues, including case selection, operational premises, data collection, data analysis strategies, and methodological implications.

While environmental threats are far from new (Beck, 1992; Carson, 1962; Steingraber, 1998), we have but a limited insight on community-level decision-making processes in spite of our need for such knowledge. Hence, the strategy employed for examining these issues must be capable of unpacking poorly understood social relationships between industry, community, and physical environments, a challenge well suited for qualitative research (Alasuutari, 1995). To accommodate these needs, this project utilizes specific case selection criteria and operational premises. In this instance, the operational premises serve as a guide to focus data collection and analysis efforts rather than lay the foundation for a strictly empirical, hypothesis-driven, test. Data collected from both historical documents as well as in-depth qualitative interviews that permit key informants to contribute new perspectives about the phenomena under study pave the way for expanding our limited knowledge about community-level responses to long-term environmental threats.

Case Selection

The focus of this project is to better understand the processes that environmental regulators, industrial entities and community members engage in as they define significant environmental threats and related mitigation actions, and how such processes impact community cohesion. Thus, the goal of case selection was to identify communities associated with documented environmental threats that pose comparable public health risks. An important component of public health risk definitions are the physical characteristics of risks that may potentially cause harm to human and ecological health (Finkel and Golding, 1994; Wing, 1994). Matching criteria considered for such physical parameters include: 1) chemical composition of the contaminants; 2) duration in which contaminants are present; 3) potential exposure pathways and health effects; 4) amount of land affected; and 5) geological concerns of the affected land. This study examines community-level decision-making rather than differences in how federal agencies manage environmental risks across regional districts so this study further limited case selection to those within a common federal district.

Since the Superfund process defines environmental threats, making information available that is otherwise difficult to obtain or nonexistent, Superfund sites considered for selection were those within Environmental Protection Agency (EPA) Region 10 (the states Alaska, Idaho, Oregon and Washington). Among those four states, only Oregon and Washington fall within the same Bureau of Land Management (BLM) district. In order to assure the selected cases would be managed by the same federal districts, case selection was further limited to Oregon and Washington.

Different classifications of National Priority List (NPL) sites also need to be taken into account. The four NPL categories that the EPA assigns to sites are: 1) proposed for preliminary site evaluation and hazard ranking but not yet formally placed on the final NPL list; 2) hazard ranking score greater than 28.5 (on a scale of 0 for no risk to 100 for maximum risk) at the conclusion of the preliminary site evaluation and significant enough of a public health risk to warrant formal placement on the NPL; 3) deleted from NPL list after completion of site remediation; and 4) proposed but removed from NPL consideration due to minimal public health concerns or management by another regulatory program (EPA 1992). Sites meeting formal NPL listing criteria (categories 2 and 3) have the most information available and as determined by the EPA, pose public health threats significant enough to warrant federal mitigation (EPA 2000d). The distribution of NPL sites in Oregon and Washington according to NPL status is as follows:

Table 2.1 Distribution of NPL Sites in Oregon and Washington by Status

	Oregon	Washington
Proposed for NPL	2	2
NPL Final List	10	56
Deleted from NPL	3	14
Removed from NPL	3	7

At two of these NPL sites, approximately 25 years of mining and metal manufacturing processes left behind radionuclides, heavy metals, and mixed solvents–the most ecologically toxic and highest priority contaminants with respect to welfare risks on the federal policy agenda (Finkel and Golding, 1994). Both sites involve a similar amount of affected land and share common geological concerns. Exposure pathways of greatest concern at both sites are surface and ground water. However, contaminants of significance are also present in soil and air pathways. While the public health threats to the communities associated with these wastes are similar, mitigation responses to each case are quite different.

Teledyne Wah Chang Albany

The first site is located in Millersburg, Oregon, about 65 miles south of Portland, Oregon. The Teledyne Wah Chang Albany[1] (TWCA) facility produces metal alloys, including zirconium, a material used in the construction of fuel rods for nuclear reactors. The area designated as a Superfund site consists of 285 acres where the old plant operated from 1957 to 1980 and an abandoned disposal area for sludges containing radionuclides, heavy metals, and chlorinated solvents. On December 30, 1982, this site was proposed for the National Priorities List but not officially listed until September 8, 1983. The site borders the Willamette River, Truax Creek and Murder Creek. These waterways support recreation activities, fishing, watering of livestock and irrigation. About 1,100 TWCA employees continue to produce zirconium and metal alloys nearby. Several other industrial operations are active in the area including a paper mill, several packing plants and distribution centers, and a railroad yard but TWCA is the only Superfund site in the area. Approximately 30,000 Millersburg and Albany, Oregon residents live within three miles of the site. Hence, with respect to environmental threats and their mitigation, Millersburg and Albany residents share interests in waste management activities (EPA 1989, 1994, 1995). Mitigation of affected sediment and soil concluded in 1998 and 1999, respectively. Cleanup actions for groundwater are ongoing.

Dawn Mining Company

The second site is located eight miles from Wellpinit, Washington, centrally located on the Spokane Indian Reservation and about 45 miles northwest of

Spokane. From 1955 to 1981, the Dawn Mining Company (DMC) operated a 320 acre open-pit uranium mine on the site. In addition to on-site sources, seep waters from the mine, containing radionuclides and heavy metals, feed drainages that lead into Blue Creek. Blue Creek flows into the Spokane River arm of Lake Roosevelt, a popular subsistence fishing and recreation area. According to the Spokane Tribe, 1,230 tribal members live on the Spokane Indian Reservation, almost all of whom live within a 17-mile radius of the site. Moreover, 140 tribal members live along the mine's hauling route (EPA 2000d). This site was proposed to the NPL on February 16, 1999 and officially listed May 17, 2000. Site characterization is ongoing, leaving room for the identification of additional contaminants (Bureau of Land Management 1996a, 1996b, 1996c).

The Dawn Mining Company also owns 820 acres near Ford, Washington, 18 miles away from Midnite Mine. Between 1956 and 1982, the DMC used 147 of the 820 acres to mill uranium ore extracted from the Midnite Mine. Ecological contaminants at this site include radionuclides, heavy metals, silica complexes, sulfates, phosphates, and chlorides. The property boarders Chamokane Creek which flows into the Spokane River, and is adjacent to, but not on, the Spokane Indian Reservation. The Spokane Tribe of Indians executive order that stipulates the reservation boundaries includes Chamokane Creek and the tribe holds water right to the creek as well. Approximately 500 people live within six miles of the site. Currently, the mill site is not part of the Midnite Mine NPL site. Instead, when the mill closed in 1982, the state of Washington drew upon the National Environmental Protection Act (NEPA) and the state version of NEPA to negotiate mitigation actions. Under the guidance of the Washington State Department of Health, mitigation of the mill site is ongoing (Washington State Department of Health 1991).

Potential impacts of both the Midnite Mine and DMC mill site extend to the communities of Wellpinit and Ford, Washington. While the State of Washington and the EPA treat these sites and communities as separate entities, the residents, the Natural Resources Trustee Council, and the Department of Justice frame it as one community and one site. Not only do the local residents share interests in both sites but they participate in activities concerning the management of the mine and mill largely as one group. Similarly, while the state and federal authorities involved with both the Midinite Mine and the DMC mill site recognize differences in lead authority, they frequently interact with each other as the management of both sites is highly interrelated. Hence, this study treats the Dawn Mining Company's Midnite Mine and mill site as one interrelated case and will examine how the communities of Wellpinit and Ford, Washington, perceive and respond to both the Midnite Mine and DMC mill site. Historically and presently, Spokane Tribe of Indians is a subsistence-based population that practice traditional cultural lifestyles that rely heavily on natural resources. Employment supplements this lifestyle to varying degree and centers on natural resource extraction, including the reclaimed Sherwood Uranium Mine and Mill Site on the Spokane Reservation. Currently, various logging operations, farming and a casino provide local employment, with natural resource and facilities management, and tribal administration, providing the most jobs. Residents also commute to Spokane for work and other amenities.

Summary

The communities impacted by the TWCA and DMC Superfund sites (Millersburg and Albany Oregon, and Wellpinint and Ford Washington, respectively) share important physical contamination and environmental management concerns with respect to exposure pathways. Both cases involve threats to surface and ground water from radionuclides, heavy metals and solvents resulting from nearly 25 years of industrial activities. However, social differences (i.e. small urban, rural and tribal settings) may produce a wide range of community risk definitions and responses to similar environmental threats. Different mitigation responses may also be indicative of other community-level social and cultural differences not currently understood but of interest to this project. Expanding our understanding of this range of responses will enhance our ability to plan and manage mitigation decisions elsewhere.

Operational Premises

As stated earlier, the focus of this investigation is to better understand the processes that decision-makers (i.e., environmental regulatory authorities) and community members engage in as they define significant environmental threats and related mitigation actions, and how such processes impact community cohesion. Here, the concept of community cohesion refers to the magnitude or degree to which a structured framework of social and biophysical characteristics exists, that enables community members to sustain a common life, and form collective responses to environmental threats. As discussed in chapter one, the social construction, disaster, risk and community theory literatures suggest the following characteristics are important for developing and maintaining community cohesion: 1) shared history; 2) community identity (e.g., geographical boundaries, historical images, physical structures, stigma effects, and attachment to place); 3) control in local decisions; 4) distribution of power among local institutions; and 5) participation in decisions about environmental threats and mitigation. Similarly, the Environmental Protection Agency's policies for public participation and consultation direct agency representatives engaged in Superfund site management activities to know the affected parties, involve affected parties as soon as possible, and establish ongoing relationships with affected parties built on honesty and integrity (EPA 2000b, National Environmental Justice Advisory Council 2000). Hence, the manner in which environmental regulatory agency personnel and community members utilize these principles when they define an affected community may shape both the form of mitigation efforts and ownership for mitigation decisions. More specifically, the operational premises, or the proposed role of community attributes in community-level decision-making about environmental risks, for this study are:

Operational Premise 1: The more uniform the ideas environmental regulators, industrial entities and community members hold about affected parties, the more likely they will reach consensus about the significance of environmental risks and need for mitigation.

Operational Premise 2: The more uniform the ideas environmental regulators, industrial entities and community members hold about affected parties and public health risks, the more likely it is that the affected community will cohesively support mitigation decisions.

Operational Premise 3: A decrease in consensus among environmental regulators, industrial entities and community members will produce an increase in cohesion among community subgroups organized along lines of conflict, and a decrease in community-wide cohesion.

Increasing our understanding of these important social relationships may better equip both decision-makers and communities to effectively manage environmental risks.

Data Collection

The two principal data sources for this study are: 1) historical documents; and 2) in-depth qualitative interviews with environmental regulatory agency personnel and community members, i.e., key informants, involved in the characterization of environmental threats and mitigation decisions. Historical documents provide background information about each case (e.g., contaminants and industries involved, associated perceived risks, affected water, soil and air resources, amount of area affected, mitigation intents, risk communication activities), and aid the identification of key informants for initial interviews. Interview data will help fill gaps present in the documents and identify additional information needed for effective decision-making. Both of these data sources explore the nature of the perceived threat, the mitigation process as it unfolds, and associated changes in critical features of community. A detailed description of these two data sources follows.

Historical Documents

The historical documents consulted in this study comprise a wide range of sources. First, it was necessary to identify the cultural context that led to the establishment of the potentially responsible parties, or companies involved in the creation of these hazardous sites, as fixtures in the affected communities. This required unraveling historical accounts about the transformation and development of the Oregon Territory with respect to the formation of different land use beliefs and community identities. Specific topics investigated and described in chapter four include the indigenous cultures of the Oregon Territory, congressional acts promoting settlement predominantly by whites, colonization, and patterns of land use and industrial development.

In order to describe the documented history of concerns associated with the TWCA and DMC sites, I reviewed Records of Decision, site characterization reports, and mitigation plans prepared by the EPA and BLM. Additional documents obtained from the EPA's administrative files through the Freedom of Information Act include hazard rating scores, site inspection reports, site descriptions and fact sheets, public involvement plans, public meeting transcripts, and public comments. Environmental Impact Statements prepared by the Washington Department of Health provided information about the DMC mill site operations as well as public comments about site management activities. Altogether, these documents provided feedback from at least 158 people interested in the TWCA site and 212 people interested in the DMC mine and mill sites, in the form of comments made at public meeting and/or written statements. However, not all of the commentators were identifiable nor was age and gender information available for all persons that contributed to the historical documents. Nonetheless, these historical data furnish critical information about the TWCA and DMC sites (e.g., contaminants and industries involved, associated perceived risks, affected water, soil and air resources, amount of area affected, mitigation intents, risk communication activities), and provide a valuable supplement to key informant interviews. In addition, articles from local newspapers (i.e., *Albany Democrat-Herald, Corvallis Gazette-Times, Lebanon Express, Oregonian, Wellpinit Rawhide Press, Spokesman Review*) about the sites appended perspectives of environmental regulatory agency personnel and community representatives.

Key Informant Interviews

The initial sampling frame comprised of key environmental regulators, industrial entities and community members involved in site assessment and mitigation activities as identified in historical documents. In addition, this study utilized snowball sampling techniques such that I asked informants for names of other key persons involved in making decisions about the sites during interviews. Those key informants who consented to interviews were in turn, asked to name others involved in the decision-making processes. This continued until the investigator identified no additional key informants. Other interview questions address interests environmental regulatory agency personnel and community members represent, definitions of affected parties, knowledge about affected parties, and involvement in decision-making processes (Appendix A).

Key informant interviews presented many challenges. Given the perceived sensitive nature of the issues, general distrust of outsiders, and the millions of dollars involved in mitigation actions (i.e., an estimated $160 million for the

Midnite Mine), it took about 12 months to foster an interest in the study among Wellpinit and Ford, Washington key informants. In addition, shortly after initiating and completing interviews with four Oregon key informants, the TWCA facility entered into a labor strike that unexpectedly lasted seven months. Until the strike concluded, the remaining Oregon key informants were not comfortable discussing any of the issues. All interviews with Oregon and Washington key informants, however, took place between October, 2001 and August, 2002.

In order to meet the needs of respondents, the interviews were conducted in a variety of formats. All of the key informants requested to review the questions before consenting to interviews. Before scheduling the interview, one respondent requested to have time to review potential answers with legal counsel. Three key informants requested phone interviews to protect their identity. To accommodate schedules of other key informants, an additional eighteen interviews took place over the phone. After brief conversations, three key informants preferred to submit their answers via email as a result of having been misquoted multiple times in the past. While five Oregon key informants and twelve Washington key informants consented to face-to-face interviews, only one consented to tape recording the interview. Thus, I took detailed notes during all interviews.

There are additional reasons justifying the employment of a mixture of methods. The nature of the issues, economic burdens, and strongly litigative context associated with NPL sites require design flexibility in order to uncover the true manner in which these decisions take place. Furthermore, in spite of these multiple interview formats, response patterns were clearly identifiable and very consistent with historical documents. Key informants also provided valuable insights about how to interpret historical documents and offered undocumented information. Flexibility in collecting the interview data unequivocally broadened the understanding of the issues under investigation well beyond what the historical documents alone furnished.

Respondents

Altogether, twenty-one key informants (9 females, 12 males) involved with the TWCA site participated in interviews lasting from 15 minutes to 2.5 hours. The average interview length was slightly more than one hour. Five of the TWCA key informants are environmental regulators and the remaining sixteen participants are community members and industrial representatives. Twenty key informants (10 females, 10 males) involved with the DMC mine and mill site participated in interviews lasting 15 minutes to 5 hours, with an average of about 1.5 hours. Six of the DMC key informants are environmental regulators, six are tribal representatives, and the remaining eight participants are community members and industrial representatives. Maintaining the confidentiality of the key informants' identities is crucial. Since the environmental threats involved in these cases will persist for indeterminate periods of time, the decision-making processes extend across multiple decades. Environmental regulatory agency personnel need to maintain cooperative working relationships with each other, industrial entities and community members for the processes to be successful. Thus, additional details about key informants will not be provided.

Nonrespondents

In addition to the key informants who participated in interviews, 11 (5 females, 6 males) Oregon key informants did not participate. Two individuals referred me to other agency staff more familiar with the site and one referred me directly to TWCA. Three people felt their involvement in the decision-making process was too inadequate to comment on it, while two others simply did not have time. The remaining three potential informants cited other reasons for not participating such as:

> The paper mill is what really smells, not [Teledyne Wah Chang Albany]. . . .I've lived here for 25 years and there aren't any problems that I'm aware of . . . I couldn't tell you about what people think.
>
> It's a nonissue so far as we can tell. . . . I haven't heard anything in our paper for years.
>
> There were some concerns 10 years ago when they were moving the sludge but that's about it. I attend school board meetings and other community organization meetings and it never comes up.

Similarly, four Washington key informants declined to participate. They explained their nonparticipation by stating that they never really got very involved and did not have anything to say about the mine or mill site. In support of those community members involved with the DMC site, one said:

> I didn't get very involved in all that. We really trusted the group that represented the community and are very proud of all their hard work [tearfully]. They spent hours and hours of their own time on it.

Data Analysis

After converting data gathered from in-depth qualitative interviews into a computerized format, data were subjected to inductive analyses employing a variety of sorting and keyword searches to identify response patterns. These processes allow the analyst to aggregate all of the statements concerning a given phenomenon in order to construct finer-grained examinations from which specific themes emerge (Alasuutari 1995, Weller and Romney 1988). As appropriate, quantifiable descriptive statistics accompany qualitative findings. Historical documents provided a context for qualitative data interpretations. Discussions of central themes identified by the inductive analyses follow in the proceeding chapters.

Note

1   The Wah Chang company, a metal alloy producer, took over the United States Bureau of Mines zirconium production in Albany, Oregon, in 1956. In 1967, the Teledyne corporation purchased Wah Chang, renaming the company to Teledyne Wah Chang. In 1972, the company modified its name to Teledyne Wah Chang Albany (Darst et al. 1979). Teledyne Wah Chang Albany (TWCA) is the company's formal name in most Superfund documents, and thus, is the acronym used in this project. As a company spokesman explained, however, Oremet, another company that produces titanium in southwest Albany, and TWCA were merged when TWCA's parent company, Allegheny Teledyne, purchased Oremet in 1998. While the company name resulting from the merger is now Oremet-Wah Chang, the two facilities remain physically separate with Oremet to the south and Wah Chang to the north. During personal interviews, key informants referred to the company by all four names, with Wah Chang being the most common among local residents.

Chapter 3

Defining Environmental Risks:
An Overview of the Superfund Process

Introduction

Decisions about how to best manage environmental risks begin with defining substances present in the environment as hazardous in some way. In conjunction with social processes, we make claims of risks as we interact with, interpret and attach meanings to ecological processes (Hewitt, 1998; Hannigan, 1995). As discussed in chapter one, the social amplification of risk model proposes that this can take place through a variety of mechanisms including personal experiences, experiences shared by family, friends and others, and participating in decision-making processes related to said risks. However, we seldom, if ever, acquire complete knowledge about human-environment interactions. This frequently leads to competing interpretations and lengthy debates about the significance of environmental risks (Anderson, 1997; Hannigan, 1995; Rosa, 1998). That is not to say only those risks that we define as problematic are cause for harm. Environmental risks that we do not have the means to see, smell, taste, quantitatively identify, or completely understand continue to impact both human and ecological health in unexpected ways (Beck, 1992, 1995; Carson, 1962; Erikson, 1994; Steingraber, 1998). Furthermore, not all groups and individuals have an equal opportunity to engage in such critiques (Winner 1993).

Oftentimes, only those risks socially deemed legitimate capture our widespread support and are the object of organized responses that target their minimization, elimination, or management (Hannigan 1995). Within this framework, there are several environmental regulatory mechanisms that play a role in legitimizing risks. The first legislation to specifically address abandoned and uncontrolled wastes, however, was the Comprehensive Environmental Response, Compensation, and Liability Act of 1980 managed the Environmental Protection Agency, more commonly known as Superfund. Currently, the Comprehensive Environmental Response, Compensation, and Liability Information System (CERCLIS) developed for the Superfund program, maintains information on over 42,000 potentially hazardous sites in the United States (EPA 2003a). The CERCLIS sites determined to be the most hazardous form the National Priorities List. There are presently 1,235 sites on the NPL with an additional 59 proposed, 216 deleted, and 19 partially deleted, for a total of 1,529 sites, representing less than 5% of all CERCLIS sites (EPA 2000a).

In order to classify a site for inclusion on the NPL, a threat to environmental and public health must be evident and the magnitude of the threat must be severe enough to warrant federal intervention (EPA 1994a). Severity in this context is determined by a series of preliminary assessment procedures and hazard rankings. Severity alone, however, does not dictate the type of mitigation actions that will be performed at any given site. Other factors considered include the size of the population affected, exposure opportunities, feasibility and cost of implementing control measures, and state and community acceptance of control measures. This chapter will examine the processes employed that designate potentially hazardous sites as national priorities. First I will describe how a potential environmental concern becomes a CERCLIS site. Then I will explain the scoring and ranking process that may promote a CERCLIS site to NPL status. Finally, I will discuss policy goals for involving the public in these processes.

Preliminary Assessment Petitions

Under the Environmental Protection Agency's (EPA) National Contingency Program, any person may report a suspected hazardous waste site to the National Response Center Hotline, 24-hours-a-day, seven-days-a-week (EPA, 1988, 2000a). In the case of emergencies, such as spills from train derailments, truck accidents or industrial plant incidents, the EPA, state and/or local authorities work together to formulate a response as best determined by the specifics of the spill and jurisdiction in which the spill occurred. In many emergency situations, programs outside of Superfund are better able to address immediate intervention needs but Superfund resources may be appropriate for the management of long-term concerns in some instances. For sites determined not to be emergencies, any person can petition the EPA to investigate a potentially hazardous situation. Petitions are not anonymous and must include the following information: 1) full name, address and phone number of the petitioner; 2) a description of the location of the release; 3) details about how the petitioner is or may be affected by the release; 4) the type of substances released if known; 5) activities taking place at the location of the release; and 6) state and local authorities contacted about the release (EPA, 1988).

During the petition review process, designated on-scene coordinators work in conjunction with designated regional and national response teams to determine the appropriate lead agency to follow-up on the reported concern, taking into account the type of release and jurisdiction in which the release occurred. For example, Superfund does not cover petroleum or natural gas releases, engine exhaust emissions, normal use of fertilizers or pesticides, certain releases in the workplace, and some releases of nuclear materials. These types of releases fall under other federal programs (EPA, 1992, 2000a). Briefly, on-scene coordinators direct and coordinate all efforts at the site of the suspected release. They are responsible for informing all appropriate public and private interests, recommending lead agency actions, and initiating preliminary assessments. Regional response teams, comprised of federal agency, state, local, and where relevant, tribal government representatives, provide advice to on-scene coordinators and determine lead agency actions. Regional response teams can also seek advice from the national response team, comprised of designated federal agency representatives, when they are not able to resolve issues among themselves. Once determined, the appropriate lead agency has one year from the receipt of the completed petition to conduct a preliminary assessment of the suspected hazardous site, if an investigation has not been previously performed. If the lead agency determines a preliminary assessment is not necessary, the lead agency must provide the petitioner with an explanation (EPA, 1988).

The purpose of the preliminary assessment is to determine the type, magnitude, and severity of the suspected release. To fulfill this purpose, on-scene coordinators gather information describing the release, attempt to determine who the parties responsible for the release are, and assess the feasibility of its removal. Where ever possible, on-scene coordinators encourage responsible parties to remove the release and oversee removal actions. Upon completing this phase of evaluation, on-scene coordinators submit a preliminary assessment report to the regional and national response team, specifying one of the following determinations: 1) no further action is necessary; 2) management of the site falls under non-Superfund programs; or 3) conditions at the site warrant further investigation under Superfund, including the completion of the Hazard Ranking System process (EPA, 1992, 2000a).

Hazard Ranking System

The Hazard Ranking System (HRS) is a numerically based screening process employed by the EPA to evaluate preliminary assessment and inspection data on sites proposed for inclusion on the National Priorities List. The goal of the original NPL published on September 8, 1983 was to identify at least 400 sites posing a significant threat to human health and/or the environment. By setting the HRS evaluation score to 28.5 or above (on a scale of 0 for no risk to 100 for maximum risk), EPA identified 406 sites for the original NPL. There are now over 1,500 sites on the NPL. The HRS score continues to serve as an NPL status guideline, however, the HRS score alone does not determine how or when an appropriate response to a particular release will take place. More detailed site investigations typically follow listing a site on the NPL in order to better understand and manage its specific risks (EPA, 1992, 2000a, 2000d). For this reason, the EPA describes the NPL as primarily an information tool for identifying 'facilities and sites or other releases which appear to warrant remedial actions' (EPA, 2000d, p. viii). More specifically, the EPA states that:

Inclusion of a facility or site on the list does not in itself reflect a judgment of the activities of its owner or operator, it does not require those persons to undertake any action, nor does it assign liability to any person. Subsequent government actions will be necessary in order to do so, and these actions will be attended by all appropriate procedural safeguards (EPA, 2000d, p. viii).

The HRS scoring process considers three categories of factors: 1) the likelihood of the site to release hazardous substances into the environment; 2) the toxicity and quantity of wastes; and 3) number of persons and types of sensitive environments potentially affected by the release of hazardous wastes (EPA, 2000c). The four pathways or routes of potential exposure scored for potential threats are ground water migration (S_{gw}), surface water migration (S_{sw}), soil exposure (S_s) and air migration (S_a). Each pathway may receive a maximum score of 100. Scores for each pathway are combined into an overall site score using the following root-mean-square formula and range from zero to 100:

$$S = \sqrt{\frac{S_{gw}^2 + S_{sw}^2 + S_s^2 + S_a^2}{4}}$$

When scores for each pathway are low, the overall HRS score will be low. However, the overall HRS score may be relatively high if only one pathway score is high. Initial HRS scores are also subject to public review and comment, and this may result in some score modifications as additional information becomes available.

Independent of the HRS score, two other mechanisms authorize the placement of a site on the NPL. Each state or territory designates one site of highest priority under the federal order, 40 CFR 300.425(c)(2). Secondly, federal order 40 CFR 300.425(c)(3) authorizes the inclusion of sites on the NPL that meet all three of the following criteria: 1) the Agency for Toxic Substances and Disease Registry issues a health advisory for the site; 2) EPA determines the site poses a significant threat to public health; and 3) EPA determines it is more cost-effective to manage the site through Superfund than through other procedures (EPA, 2000d, p. x). Like the HRS score, the listing of a site under both of these mechanisms is also subject to public review and comment.

The highest possible score a site can receive is 100. To date, 28 sites have HRS scores exceeding 70 with the highest HRS score ever assigned being 90. The HRS score of 90 belongs to the remote Triumph Mine Tailings Piles in Blaine County, Idaho. Heavy metals, including lead and arsenic, have migrated from the waste piles at this 60-acre inactive mine into residential soil, ambient air and nearby wetlands (EPA, 2002a). While proposed to the NPL in 1993, this site has never been formally listed or mitigated, largely due to its perceived isolated location. The second highest HRS score, 87, belongs to the Murray Smelter, a former lead smelter that operated for 77 years in Salt Lake County, Utah. Lead and arsenic affected soil, ground water, surface water and sediments at the 142-acre site and surrounding area. The majority of clean-up work for this site was completed in 2001 (EPA, 2002b). The third highest HRS Score is 85 and belongs to the Big River Mine Tailings/St. Joe Minerals Corp. site in Desloge, Missouri, about 70 miles south of St. Louis in a region often referred to as the 'Old Lead Belt.' St. Joe Minerals Corp. operated the site, disposing lead-, cadmium-, and zinc-rich mine tailings over 600 acres between 1929 and 1958. In 1977, heavy rains caused approximately 50,000 cubic yards of the tailings to slump into Big River, significantly impacting aquatic life. The site, however, was not formally listed until 1992. At this time, remediation activities are ongoing (EPA, 2002c).

One of the best-known Superfund sites and the site that established the Superfund Program, is Love Canal, the 70-acre Hooker Chemicals and Plastics landfill in New York State. Love Canal received a HRS score of 54. Remediation construction activities at this site concluded in 1999 (EPA, 2000d). In comparison, the Summitville Mine in Rio County, Colorado where open pit mining and cyanide heap leaching recovered gold as well as caused numerous fish kills, received a HRS score of 50 when formally listed in 1994. Mitigation at this site is ongoing (EPA, 2002e). Times Beach, Missouri, where waste oil containing dioxin was applied to roads for dust control, received a HRS score of 40. This site became a state park following the completion of remediation in 1997 (EPA, 2002f). As all of these examples indicate, there is no clear, uniform response to HRS scores by the EPA. Instead, EPA mitigation responses are formulated on a case-by-case basis. Hence, while an HRS score provides guidance for placing a site on the NPL, it does not prescribe a course of action for mitigation.

Public Participation in Superfund Decision-making Activities

Following the formal placement of a site on the NPL, the EPA initiates a thorough site characterization, or detailed investigation, of all the potential contaminants and routes of exposure present. Different methods utilized to gather data about potential risks and exposure pathways include ecological investigations, public meetings, formal and informal interviews, and surveys. The EPA also develops proposed plans for mitigation activities. In part, this process requires the EPA to specify strategies for informing potentially impacted parties of Superfund activities by developing a Community Involvement Plan tailored for each site under the guidance of EPA's Public Participation Policy (EPA, 1981, 2000c). Although not formalized until 1981, the EPA's Public Participation Policy fulfils the 1946 Administrative Procedure Act which requires all federal agencies to hold a public meeting before implementing new actions or changes in policies (EPA, 1994a; Williams, 2002). With respect to Superfund, this means publishing the proposal to place a site on the NPL, the final listing of a site on the NPL, proposed mitigation plan, and final mitigation plans, in the *Federal Register*. Additional notification activities are completed at the discretion of the lead agency.

Specific goals of the EPA's general policy for public involvement include: 1) to foster a spirit of mutual trust, confidence and openness between the agency and the public; 2) to fulfill legal requirements imposed by environmental statutes; 3) to ensure that the agency consults with interested or affected segments of the public and takes public viewpoints into consideration when making decisions; 4) to ensure that the agency provides the public with information in a time and form that it needs in order to participate in a meaningful way; 5) to learn from, and solicit assistance from, the public in understanding consequences of potential hazards and their management; 6) to keep the public informed about significant changes; 7) to foster, to the extent possible, equal and open access to the regulatory process for all interested and affected parties; 8) to ensure the public understands official programs and the implications of mitigation; and 9) to anticipate conflicts and encourage early discussions of differences among parties (EPA, 1981, 2000b). These recommended activities are a topic of recent discussion, launched by a ten-day, on-line forum, initiated on July 10, 2001. While this forum focused on public participation with respect to all EPA programs in general, specific items pertaining to Superfund were involving local communities and making risk communication more culturally appropriate (EPA, 2001a). In addition, new policies resulting from these activities recommend expanding community involvement beyond mandated notification of NPL designation and mitigation plans to include increasing the agency's awareness of public risk perceptions, improving risk communication efforts, involving affected communities in site assessments, and gathering more comprehensive community information. The extent to which agency personnel carry out these activities and specific community-level attributes that agency personnel should acquire data on, however, remains discretionary (EPA, 2002g).

Community Involvement Plans stipulate a process for making Superfund documents publicly available at an agreed upon and generally accessible location within the affected community, like a local library. Another commonly recommended practice is to advertise notices of periodic public meetings in local newspapers and other means suitable for individual communities. The EPA may also distribute fact sheets about key stages of characterization and mitigation plans throughout a community. Sometimes the EPA forms task forces or panels comprised of community members and stakeholders as a means of gathering and sharing information. Other strategies may include review groups, workshops, conferences, monthly newsletters, public service announcements, news releases, and educational programs (EPA, 1981). All the while, full decision-making authority remains in the reigns of the EPA. Decisions are not voted on throughout the process. Said another way, the EPA seeks public input and shares information but retains all decision-making authority (EPA, 2000b). Public involvement begins and ends with information exchange. Interested community groups may apply for funding to better enable their participation in information exchanges through the EPA Technical Assistance Grants (TAG) program. With this funding, communities may hire independent experts to help them become more knowledgeable about the hazards present and technologies that target hazard management. At this time, however, TAG funds are limited to $50,000 per site (EPA, 2003b).

Summary

This chapter provides an outline of the decision-making processes and goals for the Superfund program. As this overview suggests, decision-makers attempt to tailor investigation and mitigation strategies to specific site needs but community involvement in local decisions may be minimal and inadequate from the affected parties' perspective. This decision-making framework, however, provides a mechanism for understanding how the structure of decision-making processes might impact community cohesion and the ability to formulate feasible and timely community responses to environmental risks (Eyles, 1997). As Reich proposes, 'the effective scientist in the area of public policy thus must know about the limits of scientific uncertainty, the demand for certainty by bureaucratic organizations, and the consequences of social conflict' (1983, p. 320). The case studies that follow will evaluate such consequences more specifically.

Chapter 4

This Land is Whose Land?
The Role of Shared History and
Community Identity
in Environmental Decision-making

Kind of Immigrants Wanted. To the intending immigrant who has scarcely a means to pay his way here and trusting luck to make a living after he arrives, Linn County makes no inducements. . . . Those too, with little or no capital, who do not enjoy good health, both in body and mind, had better remain in their own country and among their friends (Alley, 1889, p. 31).

We've lived among them all our lives, yet they don't know anything about us. They've never wanted to know anything about us, and I think that I opened a lot of eyes right in our own community. Yeah, I think they are beginning to realize that we're people. They just tolerated us, but they are beginning to realize that we have some things to offer and maybe some are going to be . . . wanting to know more about people, us people down here. The first people that were here (Anonymous Elder of the Spokane Tribe interviewed by Brian Flett, 1997).

Introduction

Briefly recapping the argument made in chapter one, examining the historical evolution of human interactions with physical environments is essential for decision-makers for a number of reasons. History may provide a sense of rootedness, belonging, and commitment to community members (Selznick, 1992). In turn, sharing history creates an opportunity to develop a dialogue that establishes and justifies beliefs (Cottrell, 1977), formulates ideas about community identity (Cohen, 1985; Filkins at al., 2000; Kaufman, 1977; Lynch, 1988; Selznick, 1992), and provides a framework for making decisions about what activities produce public health risks (Hannigan, 1995; Slezak, 1994). Understanding such dialogues can enhance decision-makers' awareness of different risk perceptions and underlying conflicts that may deter agreement about mitigation actions. Knowledge about shared experiences may also advance decision-makers' efforts to properly identify affected parties and the magnitude of effects. Finally, many environmental policies require the consideration of potential impacts to historical resources. In this sense, understanding an community's shared history and identity provides an opportunity to learn from past mistakes and improve collective responses to future risks.

In order to understand the decision-making atmosphere of the communities under study, one must examine how the companies deemed responsible for generating the environmental risks currently of concern became community entities. Albany and Millersburg lie within the Willamette Valley of Oregon whereas the Spokane Valley of Washington houses Ford and Wellpinit. Both the Willamette and Spokane valleys, formerly contained within the Oregon Territory, promised to secure the markets of the Pacific and Indian Oceans in an era when expansion of Mississippi Valley resources and goods seemed limitless (Clark, 1927). Securing those markets, however, displaced numerous indigenous tribes and involved a shift in communal land use beliefs to individualized industrial interests. The Chinookians and Kalapuyans of the Willamette Valley and the Spokanes[1] of the Spokane Valley respectively, are the indigenous groups of particular interest here. Considering that two-thirds of the uranium ore sources and over 90 percent of uranium extraction and milling processes in the Untied States reside on Native American lands (Kuletz, 1998; Schulz, 2001), like the Midnite Mine in this project, awareness of these historical transformations is especially important if one is to understand the uranium industry and its related counterparts. Historical phenomena of particular interest to the communities under study that are presented in this chapter include geography, a brief discussion of the Pre-Columbus culture and colonization, land use beliefs and industrial interests. For more details about the Pre-Columbus culture and colonization, see Appendix B.

Geography

Just as past actions shape present physical conditions, today's decisions will shape future land use options. Thus, it is important to understand how physical conditions influence decision-making. For example, the soil is much more fertile and the crop yields per acre are considerably higher in the Willamette Valley due to the nutrient rich volcanic silt on the valley floor. The Pacific Ocean coastline and the Columbia and Willamette rivers provide accessible entry routes into the Willamette Valley. The steep and frequently treacherous terrain Rocky Mountain range just east of the Spokane Valley, however, makes for a more difficult journey. The favorable conditions of the Willamette Valley promised good fortune to American pioneers as the following excerpts illustrate:

> It is a beautiful sight to behold the luxuriant wheat-fields about the last of June, just before the grains begins to ripen, and when the lovely spotted white lily--Lilium Washingtonium--stands head and shoulders higher among it, scenting all the air with its sweetness (Victor 1872, p. 200).

Never has there been a failure of crops and no farmer has yet been disappointed in his returns. . . . We have figures of a farmer who has twenty acres planted in large and small fruits and vegetables, and from this ground in the year 1888 he cleared $1,500 (Alley 1889, pp. 18-19).

Compared to the Willamette Valley, the upper Spokane Valley is more arid and rockier. These rocks are rich in many different minerals, providing an appealing location for ore exploration. The 24 metallic and 28 nonmetallic minerals identified in the area include gold, silver, copper, lead, zinc, silica (used for industrial abrasives), barium (used in oil drilling), beryllium, lime and limestone, marble and terrazzo, building stone, talc (filler for fertilizer and insecticides), peat, clays and sands (used for glass, concrete and ceramics), gas and oil (Stevens County Rural Development Planning Council, 1961; Wynecoop, 1969). However, the harsh terrain of the Spokane Valley generally encouraged white settlers to seek more appealing alternatives, such as the Willamette Valley. Thus, the less favorable conditions of the upper Spokane Valley protected the Spokanes from the rapid development that the tribes of the Willamette Valley witnessed.

In summary, compared to the Spokane Valley, the geographical conditions of the Willamette Valley provide a wider range of development options. Consequently, there may be a wider array of competing demands on the plates of Willamette Valley decision-makers than their Spokane Valley counterparts. This may increase the complexity of mitigation planning. It also reinforces the need to examine the evolution of human interactions with their respective landscapes.

Pre-Columbus Culture

Selecting a point in time to begin discussing human interactions with physical environments in the American West is not a simple task (White, 1991). Historical accounts of the West often begin with romanticized adventures of white pioneers settling land they treated as being unoccupied. This obscures the fact that over four hundred independent nations inhabited North America before the arrival of Columbus and his comrades in 1492 (Pevar, 1992). Thus, in order to understand the potential role of indigenous cultural traditions in current decision-making processes, one must consider how the relationships between indigenous inhabitants and their landscapes evolved over time. The first indigenous peoples of the Willamette Valley of Oregon and the Spokane Valley of Washington persisted for as long as 10,000 years before European encounters. These indigenous residents were generally a peaceful people that depended on natural resources within their domains for survival (Johnson, 1904; Scott, 1924). Not only was their relationship to the land the heart of their culture and livelihood, but the features of the landscape molded their culture (Clark, 1927; Mackey, 1974). A more specific examination of indigenous interactions with the Willamette and Spokane Valley landscapes follows.

Indigenous People of the Willamette Valley

There were two primary indigenous groups in the Willamette (meaning to spill or pour water) Valley of Oregon, the Upper Chinookians and Kalapuyans (Mackey, 1974). The Upper Chinookians lived at the mouth of the Willamette River where it enters the Columbia River and near the Willamette Falls, south of the Columbia River by present-day Oregon City (Clark, 1927). The Kalapuyans lived in the Willamette Valley south of the Willamette Falls (Mackey, 1974). Each Kalapuya band had its own headman and spoke the Calapooya dialect, a division within the Kalapooian linguistic stock (Mackey, 1974). Given this lingustic lineage, some call this same, 'most distinctly Oregonian' but least known, indigenous group the Calapooians (Clark, 1927, p. 42). To make matters even more confusing, observers and anthropologists link as many as 19 different names to the various bands of the Kalapuyans. Some of the confusion about different names for the Willamette Valley bands stems from the fact that white settlers named groups based on the geographic fixtures they lived by. For example, multiple sources refer to a band of the Kalapuyans living along the Calapooya River extending between present-day Albany and Brownsville in Linn County, as the Calapooya. Others ascribed names to the bands that fit within the Calapooya dialect (Clark, 1927; Mackey, 1974).

The Chinooks and Kalapuyans did not reside near large game grounds or use horses and thus, relied on small animal hides (muskrats, wood rat, geese, etc.) and cedar bark for clothing. The Chinookian diet primarily consisted of fish caught by using poles and nets, then dried. The Kalapuyans on the other hand, relied upon a vegetarian menu of dried and roasted roots like Camas and wappato, supplemented with seeds, fruits, and berries gathered from nearly 50 plants. Temporary shelters made of poles and mats provided housing compatible with the Kalapuyans' food gathering. The Chinooks built plank houses over two to three feet excavations, extending as long as 200 to 300 feet, then lined and subdivided with mats (Clark, 1927; Johnson, 1904; Mackey, 1974). Women spent their days as carriers, cooks, food gatherers, mat, hat and basket makers. The men hunted, made weapons and war when necessary. The Kalapuyans crafted particularly exquisite baskets, so tightly woven that they were capable of holding water. The Chinook constructed fine, sea-worthy canoes capable of carrying up to as many as 30 persons. Occasionally, canoes aided maritime war efforts but the under-armed tribes generally did not pose a significant threat (Bowen, 1978; Clark, 1927; Johnson, 1904). While the Chinook engaged in trade to supplement their lifestyle, the Kalapuyans rarely did (Clark, 1927; Johnson, 1904). Both the Chinooks and Kalapuyans controlled population size by practicing infanticide, especially in times of significant poverty. Both groups also buried their dead with their earthly possessions (Clark, 1927).

Indigenous People of the Spokane Valley

Aboriginal lands inhabited for centuries by the Spokanes (meaning children of the sun), an Interior Salish group, encompassed approximately three million acres in northeastern Washington, northern Idaho and western Montana. Historically, each of the three bands within the Spokane tribe (upper, middle and lower) had their own chief. The lands of the Spokane Valley affected both the social and economic life of the Spokane. In the spring, small groups dispersed from winter camps to gather food, hunt and fish. Activities of summer and early fall included root digging, and berry picking. This was also a time for many intertribal social gatherings, now known as 'Pow Wows.' In the late fall and early winter, smaller units would regroup along rivers and creeks that provided access to water and shelter during the winter months. The winter months were an important time for trading, visiting and observing ceremonies. Salmon was a principal component of their diet and a critical commodity for trading (Wynecoop, 1969).

In early times, the Spokanes supplemented their supplies by raiding other tribes. Raiding parties sought horses, food, weapons, and women for slaves. Unlike neighboring tribes, the Spokanes were typically not warlike people. They frequently formed alliances with other tribes including the Kalispels, Flatheads, and Coeur d' Alenes. This was a particularly useful arrangement for buffalo-hunting and trading expeditions. Women wore long dresses, leggings, a belt and moccasins fashioned from buffalo, elk and deer hides. Men wore buckskin shirts, leggings, a belt, breech cloth, moccasins and sometimes, fur hats. Shelters consisted of teepee frames covered with mats of tule and rectangular housing at permanent camps to accommodate larger gatherings and ceremonies. Women made mats, bags, baskets, clothing, dressed the hides, gathered fuel, dug roots and prepared meals. Men made ceremonial clothing and weapons, hunted, cared for the horses, and made war when necessary (Wynecoop, 1969).

Summary

In summary, for centuries the indigenous inhabitants of both the Spokane and Willamette valleys maintained reciprocal relationships with the land such that the land shaped their culture and their culture shaped the land. In this sense, the loss of land would cause not only physical hardship, but would also change how indigenous peoples maintain traditional ways of life. With these changes, came a loss of culture. This makes mitigation decisions not just about the loss of land deemed unsafe to human health but new land use restrictions may build upon historical land losses. Thus, understanding the effects of colonization on indigenous groups is important if we are to uncover underlying sources of conflict that may contribute to contentious decision-making.

Colonization

Colonization, the settlement and development of foreign territories by non-indigenous peoples, changed how the indigenous peoples of the Willamette and Spokane valleys interacted with their physical environments. Through a number of economic development activities and legal arrangements, land ownership shifted from indigenous inhabitants to white settlers (White, 1991). In the process, stationary lifestyles and individual land ownership replaced seasonally nomadic lifestyles within a communal land system. The new land owners viewed natural resources as assets subject to manipulation for financial gain and personal betterment. This strongly contrasted the indigenous peoples' belief that natural resources are both vehicles for survival and cultural assets (Alley, 1889; Clark, 1927; Johnson, 1904; Wynecoop, 1969). Specific mechanisms of colonization in the Willamette and Spokane valleys examined here include trading, missionary work, several laws and congressional acts encouraging white settlers to occupy the West, the removal of Native Americans to reservations, and the expansion of the railroad.

First Encounters with Whites and the Establishment of Trading

The first encounters with indigenous people of the Pacific Northwest began with a Spanish expedition, arriving in Nootka Sound on present-day Vancouver Island, British Columbia, in 1774. The first British crew arrived under the direction of Captain James Cook in 1778. His crew acquired 1,500 otter pelts from local tribes to use as winter clothing on their return voyage but instead, they sold the pelts in China for a handsome profit. The word of potential fortunes in the Pacific Northwest quickly spread, bringing numerous English and American traders to the area (Clark, 1928; Johnson, 1904). Although Spain vacated the Port of Nootka on March 23, 1795 and terminated further interests in occupying the Pacific Northwest, Britain and America continued to jointly occupy the area (Clark, 1928; Johnson, 1904).

With an eye on expanding both trade and personal fortune, fur traders began exploring additional water passages in the Puget Sound. In 1792, a group of British explorers became the first white men on record to encounter the Willamette tribes (Clark, 1928; Johnson, 1904). Some 21 years later, the North West Company established the first trading post in the area. In 1825, the Hudson Bay Company replaced the 1813 post with the construction of Fort Vancouver (Clark, 1928; Johnson, 1904). While trade routes to the Willamette Valley became more readily known in part due to a more manageable terrain, others sought an inland passage from the plains to the Pacific Ocean. In 1807, one such hopeful Canadian trapper, David Thompson of the Northwest Fur Trading Company, encountered the Spokanes and became the first white on record to enter Spokane territory. Three years later, Thompson established a trading post, referred to as the Spokane House, at the confluence of the Little Spokane and Spokane rivers. In favor of a more convenient location on the Columbia River fur trade route, the Northwest Fur Trading Company built a new post on the Upper Columbia River in 1825 and closed the Spokane House the following year (Wynecoop, 1969). All the while, Americans and British jointly occupied the area without making any land use agreements with the indigenous tribes.

Early Missionaries

As the fur trade established posts throughout the Pacific Northwest, a New England school teacher, Hall Jackson Kelley, began recruiting settlers in 1817 to spread Christianity via organized colonization (Johnson, 1904). It was not until 1834, however, that Methodist Reverend Jason Lee founded the first mission in the Willamette Valley. Four years later, two Catholic priests began missionary work at Fort Vancouver and the two faiths began competing for the Native Americans' attention. This religious feud, coupled with little conversion success largely due to the missionaries' intolerance of indigenous traditions, led to the abandonment of the Methodist mission in 1844 (Johnson, 1904; White, 1991).

Missionary work in the Spokane Valley followed a similar course. In 1819 a Catholic French Canadian couple staying at the Spokane House became the parents of the first white child born in the Spokane Valley (Jessett, 1960; Wynecoop, 1969). It was not until 1838, however, that Protestants Eells, Walker, and their families established a mission near present-day Ford, Washington. As tensions between the increasing numbers of white settlers and the tribes rose, Eells and Walker decided to abandon the Tshimakain Mission in May of 1848 (Jessett, 1960). Others did not attempt to establish missions among the Spokanes until 1866 when Chief Baptiste Peone of the Upper Spokanes requested the services of the Catholic priest Father Joseph Cataldo (Wynecoop, 1969).

Early Laws and Congressional Acts Encouraging Settlement of the West

In increasing numbers, white traders and squatters encroached upon important hunting and gathering grounds in the Willamette and Spokane valleys. As a result, the physical and cultural survival of its indigenous inhabitants became more difficult. These changing circumstances imposed the development of new relationships between the tribes and the American government. While the Americans treated the tribes as miniature sovereign nations residing within United States boundaries, Congress interpreted the United States' Constitution as justification for dictating compensation for desirable lands through treaties and forcing indigenous groups to migrate from aboriginal lands. To that end, Congress created a number of laws in 1790 'to protect Indians from non-Indians,' such as licensing trade and prohibiting non-Indians from taking Indian land without consent (Pevar, 1992, p. 3). As White (1991) suggests, the Untied States acted as if the Native Americans were minors under their guardianship. However, the United States government rarely enforced the laws.

In the Indian Removal Act of 1830, the President was authorized to negotiate the relocation of eastern tribes to the West (Pevar, 1992). Not only did this increase the conflict between white settlers and indigenous tribes, but it also placed the eastern and western tribes at odds with each other. To make matters worse, the whites brought with them a host of new diseases including measles, influenza, small pox, syphilis, and whooping cough (Cook, 1952; Taylor and Hoaglin, 1962). As a result of favorable geographical conditions spurring rapid development, about 9 out of 10 indigenous people in the Willamette Valley died from imported diseases before 1840. Thus, 'disease helped make possible the settlement of the Willamette Valley by the whites, almost without resistance, and settlement completed the collapse of the Indian culture' (Mackey, 1974, p. 21).

The first congressional act that specifically promoted colonization of the West was the Preemption Act of 1841. Under this act, American citizens or persons intending to become American citizens, i.e., a free white person, of sound moral character, residing in America for at least 14 years (Immigration and Naturalization Service, 2002), could purchase up to 160 acres for $1.25 per acre after living on it for 14 months (Bohm and Holstine, 1983). Nine years later, the Donation Act of 1850 provided public lands, full title and free of charge, to white citizens (including Americans with one indigenous parent and eligible persons declaring their intention to become American citizens), age 18 or older, residing in the Oregon Territory by December 1, 1850 (later modified to before January 1, 1855). Individuals could claim 320 acres of vacant, non-mineral land and qualified married couples could claim 640 acres (Linn County Pioneer Memorial Association, 1979; Mullen, 1971). On December 1, 1853 Congress limited claims to 160 acres for single persons and 320 acres for married couples (Linn County Pioneer Memorial Association, 1979). The only way for a Native American to keep the land he or she already occupied, was to sever tribal affiliations and become an American citizen. Many Native Americans did not understand or know about the laws nor were they interested in severing tribal affiliations. As result, they lost the lands they had inhabited for generations along with the livestock they maintained on the lands (Wynecoop, 1969).

After considerable depletion by disease, the surviving tribes of the Willamette Valley signed a treaty in 1855 that exchanged their indigenous lands for $62,260 and the Grand Ronde Indian Reservation. The reservation encompassed approximately 56,699 acres northwest of Albany, Oregon, and about 25 miles from the Pacific Ocean in Polk County. In 1858, one year after Congress ratified the treaty, a melting pot of about 1,200 Native Americans from more than 20 different tribes and bands lived on the reservation (Clark, 1927). All the while, the Willamette Valley white population soared to over 30,000 (Mullen, 1971) by 1860.

Meanwhile, acquisition of the Spokane Valley lands came with resistance. To start with, the rushed tactics of Isaac Stevens, first governor of the Washington territory, to acquire land holdings cultivated discontent among the tribes (White, 1991). Secondly, as word of the first gold strike near Colville in 1854 spread, white settlers flocked to the Spokane Valley, significantly threatening indigenous food supplies and the Spokanes' way of life. Injurious conflicts between the tribes and whites ensued. In 1858, tensions peaked when the Spokanes heard that Colonel Steptoe and 150 soliders were coming to Colville to investigate the murder of two miners. In response, the Spokanes made alliances with the Coeur d' Alenes, Yakimas, Kalispels and Palouse to defend their aboriginal lands. The alliance forced Steptoe and his troops to retreat near present-day Rosalia. Colonel Wright then led a retaliatory campaign of 700 well-armed soldiers and defeated the Spokanes. Following his victory, Wright rounded up the remaining horses of the Spokane and slaughtered them. He continued his rampage, destroying all standing crops along the way and hanging 15 Spokane for alleged murders at a site now known as Hangman's Creek. Having lost most of their food stores, many Spokane starved to death the following winter. Efforts to establish a treaty and place the remaining Spokanes onto a reservation resumed but to no avail (Wynecoop, 1969).

The third significant congressional act to encourage colonization of the West was the Homestead Act of 1862. Under the Homestead Act, American citizens and persons intending to become citizens could claim 160 acres. After improving and residing on the land for five years, those eligible could gain full title to the land for a registration fee of $26.00 (Bohm and Holstine, 1983). The Homestead Act was perhaps the single most successful colonization plan. Citizens claimed roughly 270 million acres, or ten percent of the United States, through this program (National Parks Service, 2002). By 1870, the white population of the Willamette Valley jumped to 44,488, nearly 2,000 of whom lived in the city of Albany. The Spokane indigenous lands were difficult to clear and perhaps spared the area from the rapid development seen in the Willamette Valley such that the white population of the Spokane Valley dropped from 996 in 1860 to 734 by 1870 (Bohm and Holstine, 1983; Bowen, 1974; Clark, 1927).

Spokane Reservation

While colonization of the Willamette Valley was well underway in the 1870s, agreements opening the Spokane Valley to white settlers were absent. That began to change in 1880 when the United States Army built Fort Spokane to protect white settlers from possible Native American attacks. At that time, the Spokanes were roughly 3,000 in number (Wynecoop, 1969). Less than one year later, President R.B. Hayes signed an Executive Order on 18 January 1881 that established the Spokane Indian Reservation. The boundaries of the reservation encompassed 154,602.57 acres of land not considered the most desirable for food gathering, hunting and fishing. Moreover, due to lack of payment for aboriginal lands and religious differences among the Spokane bands, the upper and middle bands refused to settle on the reservation (Ruby and Brown, 1970; Wynecoop, 1969).

A gold discovery near Coeur d' Alene in 1883, however, fueled a second significant gold rush. This increased the numbers of white settlers encroaching on Spokane aboriginal lands. In fear of losing their remaining land, the Spokanes entered an agreement on 18 March 1887 to move to the Spokane Reservation or other nearby reservations. In exchange for the move, the tribe received $127,000, earmarked for the erection of houses, purchase of cattle, seeds and farm equipment. By 1897, 340 Lower Spokanes, and 188 Upper and Middle Spokanes lived on the Spokane Reservation (Wynecoop, 1969). Establishment of the reservation did not, however, offer full protection of this small slice of aboriginal land.

Later Congressional Acts Promoting Colonization of the West

In efforts to complete colonization of the West, Congress passed the General Allotment Act in 1887. Under this act, surveyors divided reservation lands into allotments. Each tribal member age 18 or older could claim one allotment. After all eligible tribal members filed claims, the United States government made the remaining allotments available to white settlers for purchase (Tyler, 1973). This reduced the 140 million acres collectively owned by Native Americans in 1887 to 50 million acres by 1934 when Congress abolished the act (Pevar, 1992). Through this act, white settlers purchased 25,791 acres of the 56,699-acre Grand Ronde Indian Reservation (Confederated Tribes of the Grand Ronde Community, 2002). By 1920, over 200,000 whites lived in the Willamette Valley. The following year, the last Calapooian died (Stanard, 1948), just three years before all Native Americans became citizens of the United States.

Under the General Allotment Act, nearly 90,000 acres of the 154,602.57-acre Spokane Reservation were made available to white settlers. Feeding the growth frenzy, on April 9, 1910, Congress opened the Spokane Reservation for mineral leasing and extraction. White prospectors, however, failed to identify ore fortunes on the Spokane Reservation. Instead, two Spokane tribal members made the first sizable lode discovery, the Midnite Mine, nearly 45 years later (Wynecoop, 1969). Another big blow to the Spokanes' culture came with the Act of June 29, 1940 that authorized the acquisition of reservation land along the Spokane River for the construction of the Grand Coulee Dam and Reservoir. The 550 foot dam put an end to both historical salmon runs and all of the tribe's traditional ceremonies associated with them.

The final step to fully integrate Native Americans into white society was the Termination Act of 1954. This act authorized Congress to terminate all federal benefits and services offered to tribes and eliminate tribal land holdings (Pevar, 1992). Congress terminated 109 tribes, 64 of whom were native to Oregon, between 1954 and 1966, including those on the Grand Ronde Reservation (Oregon Legislative Commission on Indian Services, 1999). The political power of the Spokanes on the other hand, began to improve in the 1950's. Under the Indian Reorganization Act of 1934, the Spokane Tribe developed and approved its own constitution and bylaws on 27 June 27 1951. Less than two months later, the tribe filed a claim against the United States government for underpayment of their lands. A second docket alleged that the United States government mismanaged monies and properties that it held in trust for the tribe. The final judgement for these two claims rendered on 21 February 1967 awarded $6.7 million to the tribe (Wynecoop

1969). In 1981, additional claims alleging mismanagement of tribal assets resulted in an award of $271,431.23. The enactment of Public Law S5-240 in 1958, returned 2,752.35 acres of vacant and indisposed land declared surplus under the Donation Act to the Spokanes. While forever changed by the impacts of colonization, the tribe's 2,153 members work together to preserve their culture and to achieve self-sufficiency (Spokane Tribe of Indians, 1999).

Summary

Colonization removed all of the indigenous peoples from the Willamette Valley. Moreover, it removed a Native American voice from all land use decisions in Linn County and its towns and cities, including Millersburg and Albany. As the information provided in this section illustrates, Albany and Millersburg are the products of joint efforts to establish an industrial hub, mostly occupied and run by whites, in the Willamette Valley. The building blocks of this industrial hub include: 1) favorable geographical conditions; 2) individual land ownership where land was acquired by occupation largely in the absence of negotiation with its indigenous inhabitants; 3) transformation of indigenous land uses to place-fixed, long-term cultivation and manipulation of natural resources for economic gain and exchange; 4) marginal understanding and intolerance of indigenous traditions; and 5) insignificant resistance from indigenous groups to land transformations due to their reduction in numbers as a result of imported diseases.

The Spokane Tribe on the other hand, continues to occupy a small section of their original domain, albeit the less favorable traditional hunting and gathering grounds. Failed attempts to prevent white settlers from occupying their land resulted in an increased population density that prohibits complete adherence to traditional customs and lifestyles. At the same time, the only place the Spokanes can freely practice their traditional spiritual and eco-cultural lifeways is on the reservation. This creates an atmosphere where maintenance of culture competes with the need for survival. Not only is preserving historical connections difficult for the Spokanes amidst these shared disruptions, but incorporated into their history are imposed new roles, continuous revisions of land use agreements by the federal government without the Spokanes' consent, and the necessity to develop new relationships with their remaining land. In order to understand the role that the historical challenges and transformations of the Spokane and Willamette valleys play in current decision-making processes, however, we must also examine prevailing land use beliefs and industrial interests.

Land Use Beliefs and Industrial Interests

Beliefs about how to best use land and the resources contained within it evolve over time as humans interact with changing physical environs. These beliefs provide a framework for making decisions about what activities pose public health risks and how to best manage such risks. Hence, understanding the framework that results from cumulative interactions with physical environs is critical for mitigation planning. Dimensions of land use beliefs and industrial interests identified by key informants as being particularly important in decision-making processes and discussed here include historical preservation, acceptable industries, competing interests, decision-making time frames, and trust in decision-makers.

Historical Preservation in Millersburg and Albany

The pioneer pride of conquering western lands that originated more than 150 years ago continues to prevail in Millersburg and Albany. Visitor's brochures boast of three viable historical districts in downtown Albany and two historical museums (one run by the city and the other by the county). Included in a more contemporary historical spotlight are Linn County's 51 remaining covered bridges dating back to 1932. All of the historical landmarks of notoriety carry European origins. Outside of the two museums that house artifacts of the Willamette Valley's indigenous peoples, reference to Native American culture is not apparent. Key informants describe Albany as a 'close knit, very involved, nice small town in western America' that is 'very patriotic' and 'socially and financially conservative.' Consistent with its legacy as an early industrial center established under congressional acts favoring whites, it remains a 'blue collar town' that is 'pretty white-bread' with a 'small and growing Hispanic population, smaller black and Asian population' and 'not a diverse population so far as religion goes.' Another key informant referred to the Millersburg and Albany area as a 'red collar town,' clarifying the term to mean 'redneck, blue collar folks.' Organized community events as one key informant describes focus on American culture and are a source of local pride:

> There's no regular events in Millersburg but in Albany there's Victorian Days at Christmas where there is a tour and tea of some of the historic Victorian homes. In July there's a Timber Carnival and a Hot Air Balloon Show in August. We have one of the biggest Veterans Day parades around. People are very involved in high school sports.

Historical Preservation in Wellpinit and Ford

Conversely, the Spokanes believe everything (soil, water air, culture, people's health) is interconnected. Cultural history is extremely important and centers on relationships with the land rather than physical structures created by man. As one key informant suggested, 'creeks are the most important structures.' Today the town of Ford remains close to its original form, consisting of a post office established in 1912 with an adjoining general store, a fire station run by volunteers, and an abandoned elementary school that locals are trying to formally designate as an historical landmark. Nearby, there is a monument for the Tshimakain Mission and the St. Joseph church still stands.

While state and federal officials see Ford and Wellpinit as distinctively different geographical communities, local key informants described Wellpinit and Ford as part of one community, composed of different cultural entities. Four key informants also suggested they did not have a very good understanding of what community members' value. Other key informants described the Spokane Tribe as:

> . . . a pretty typical Native American culture, very connected to all their resources. . . . They are much more subsistence based and dependent on the ecosystem than you or I. They go beyond economic and agricultural uses of the land. There's a lot of spiritual uses of the resources. They take a much more comprehensive approach.

> They talk through things in person, they're very verbal and lots of in-person consulting goes on before making decisions.

> There's a strong attempt to maintain traditional practices, but it may differ from one family to another.

The white population on the other hand, consists of 'plain old country folks, loggers and gentlemen farmers . . . and people getting out of Spokane to live in the country.' More specifically, key informants described the white population as:

> . . . retired folks, country-type people, not nosey, and slower paced. They care about each other but don't worry about what neighbors are doing. They don't interfere and respect each other's way of life. Everybody has the right to live the way they want but we rally around each other in times of need.

> Fundamentally the people that live in the area want to be left alone. They undergo a lot of personal hardships for privacy. Instead of designer clothes with all the right labels, the social attire is a baseball cap with a pickup truck and rifle rack.

> Like every community some people are more liberal than others and some are richer than others. Some are smarter than others and some have more education. That said, I would classify my community as poorer, less literate, less involved politically, more likely to be involved in the Christian Identity movement [and support the Aryan Nations], less accepting of diversity, more likely to have been on welfare, more likely to support the 'right to bear arms,' more conservative and firmly believe that less government is better, more likely to work in an extractive industry (primarily timber) . . . than are most other communities.

Other key informants described the community' cultural mix by contrasting different groups:

There's three types of people: the people that live on the reservation; land owners that were here when [the Dawn Mining Company] came in and they grew up with [the company] harmoniously; and the new California trailer house crowd. They're the most outspoken and against everything 'come lately.' The three groups are pretty distinct.

There's a technical tribe group and a cultural group that doesn't want to disturb things, and a government group that represents the community and is concerned about land use changes.

In some ways, the rural setting and topography of the upper Spokane Valley make the area's residents 'conscientious of resource use and management.' With only 1.5% of the land within Stevens County boundaries classified as developed, it is not surprising that centralized water and sewer systems are absent. Over 60% of county lands are forest or open prairie, making forestry, agriculture, mining, and government work central to the Stevens County economy. By 1960 however, 14% of Stevens County farms were placed into the federal Soil Bank Program and 56% of farm operators worked off the farm to supplement their income. Similarly, although forestry is an important source of off-farm income, most operations are seasonal, swelling unemployment rolls on the reservation to over 80% at times (Rawhide Press, 1984). The area's many lakes, rivers, streams and parks though, provide ample public recreation opportunities including snowmobiling and Nordic skiing during the winter months (Washington Department of Health Services, 1991; Stevens County Rural Development Planning Council, 1961).

Acceptable Industries in Millersburg and Albany

With open arms, Millersburg and Albany welcome industry, and it is not clear from conversations with key informants where development limits lie. Advanced technologies entered the scene in 1941 when the United States Bureau of Mines took over the abandoned Albany College buildings to establish a branch office and experimental metallurgy laboratory. Within the laboratory's confines, in 1947 Dr. W.J. Kroll developed a process to produce zirconium, an important material used in the construction of fuel rods for nuclear reactors (Darst et al., 1979). The Carborundum Company of America took over the zirconium production for the United States Bureau of Mines in 1953 as the bureau moved ahead with other metal alloy experiments. In 1956, the Atomic Energy Commission sought bids for the production of 2,200,000 pounds of zirconium per year for five years. Wah Chang[2] won the bid, and after building a new facility, produced its first batch of zirconium sponge on Christmas Day in 1956 (Mullen, 1971). Since then, the company, now know as Teledyene Wah Chang Albany (TWCA), developed other metal alloy production processes including tantalum, tungsten, molybdenium, vanadium and titanium. Along the way TWCA transformed Albany from a mill town to a high tech mill town and became the area's largest employer, currently providing about 1,100 jobs.

Local key informants hailed TWCA as 'very cooperative,' 'a good corporate citizen,' 'provid[ing] lots of high paying jobs.' As an 'active partner in the community,' TWCA sponsors 'concerts, fun raisers, spearheaded the United Way and helped develop a world class boys and girls club.' In letters written following a public meeting, two community members were quick to recognize the area's dependence on TWCA. One even offered justification for their less than pristine environmental track record:

> The business of TWCA is the business of Albany. TWCA is the backbone of Albany's diversified economy, being our largest employer and is a responsible corporate citizen of our community. TWCA is important to our community and any unnecessary regulations and costly clean-up plans affect their ability to compete and survive which directly affects employees, citizens and the health and vitality of Albany (TWCA Administrative Record[3] 14.4-0011833).

> Teledyne Wah Chang has been a very important part of our community for the past twenty-thirty years. If there are problems that are existing it has been because there were no guide lines as there are now and when guide lines were imposed they were followed to the fullest . . .We need jobs and a good healthy economy which are being provided by Wah Chang (TWCA Administrative Record 14.0011791).

Although TWCA 'has somewhat of a reputation as a polluter,' one key informant explained, 'a lot of people are connected to [TWCA]. They either worked there or know some one who worked there, and they are very comfortable with the situation.' Said another way, 'people grew up with it and carry on.' As one key informant suggested, to outsiders 'it might seem strange, like there would be a conflict, but there isn't.' In addition to their role as the major employer, Millersburg and Albany residents think of TWCA as 'the national leader in rare metals manufacturing.' As they explained, this is a very specialized high tech industry and TWCA fills a unique world-wide niche. In letters supporting TWCA, two community members stated:

> I am pleased that the Wah Chang plant is there and that they have so many well paid and well taken care of employees. And that they can produce such necessary metals for our defense (TWCA Administrative Record 14.4-0011821).

> This is a company that cares about the environment, a company that cares about the community and its employees. This company has made a commitment to clean up what years ago was acceptable practice and forge ahead. This company should be commended for its leadership (TWCA Administrative Record 14.70004).

While one key informant reported that TWCA is 'not the only thing we run on anymore,' the economic backlash from a recent 7-month labor dispute illustrated the area's continued dependence on the company. One key informant stated that 'about five million dollars worth of purchasing power didn't happen during all that.' Offering a more specific example, another key informant stated, 'Costco sales dropped significantly during the strike and they almost closed, that's how much [TWCA] supports the local community.'

Acceptable Industries in Wellpinit and Ford

Similarly, the Dawn Mining Company 'historically has been economically important' to the Ford and Wellpinit, Washington having 'at one time supported a lot of jobs in a community that has few options.' In 1954, Spokane tribal members Jim and John LaBret discovered a fluorescent material, identified by the Bureau of Mines as meta-autunite, a uranium rich mineral. Shortly thereafter, the brothers founded the company Midnite Mines, Inc. to further investigate their discovery. Their fruitful investigation confirmed the presence of extractable ore quantities. In need of a financial backer to extract and mill the ore, a partnership between Newmont Mining Corporation (51% ownership) and Midnite Mines, Inc. (49% ownership) formed the Dawn Mining Company (DMC). Ore extraction began in 1955 following the establishment of a land lease agreement to DMC through the Bureau of Indian Affairs. Two years later, the mill produced its first batch of yellow cake uranium at the mill in Ford. Under Atomic Energy Commission contracts, DMC produced yellow cake until 1965. At that time, uranium markets weakened. While DMC did not contribute to the local infrastructure specifically, two key informants remised about a short-lived boom town near the mill site:

> Uranium City was built as a boom town right by the mill. There was a general store, supermarket, gas stations and some houses. But it's not a remote site so it really wasn't needed and that's all gone now.

Following an improvement in the uranium market and a one million-dollar renovation of the mill in 1969, DMC produced yellow cake under commercial contracts until November of 1982. Termination of activities resulted from a combination of dropping uranium prices and suspension of mining. The Spokane Tribe suspended mining in 1981 after discovering that DMC violated the extraction specifications of the lease by only removing selective high-grade ore. The extraction strategy employed by DMC at that time raised concerns about excessive erosion in some mine site areas. Although 'the tribe received between 10 and 20% in royalties and shareholders received about $7 million in dividends,' there are mixed feelings about DMC's presence. For example, key informants stated:

> Allotment owners got royalties, only when in operation, for ore on that land. Tribal members worked as miners. Three tribal members were on the DMC board. Some tribal members still work at the Ford mill. Some made a lot of money, some didn't.

> The mine was associated with economic growth that was good for the tribe. Some want to reopen it for jobs but most want it to be made safe.

Termination of DMC activities does not imply, however, that the Spokane Tribe disfavors mining altogether or uranium mining and milling in particular. In fact, from 1978 to 1984 Western Nuclear ran another low-grade uranium operation, the Sherwood Mine and Mill, named after Spokane's Chief Sherwood. Also located on the reservation, the Sherwood Mine and Mill was about four to five times larger than the Midnite Mine but only five miles from it. At the time Western Nuclear initiated this operation, 'environmental regulations were already in place . . . they planned for the closure right from the start and mined with the closure always in mind.' Hailed as 'an example of new mining,' seven key informants stated that those involved are 'very proud of its reclamation.' In fact, the reclamation was so successful that '[Western Nuclear] was 100% released from the mine and mill reclamation bond,' a first for the mining industry. Western Nuclear operated very differently from DMC and 'worked closely with the tribe all the way,' as one key informant said. Other key informants reported that:

> They brought in Robert Sherwood in the scoping phase very early and invited all tribe members. About 18 people from Western Nuclear came to the public meeting, asked for guidance from the most respected members. They asked what types of information were wanted, and provided food incentives at public meetings.

> The company [Western Nuclear] has a strong environmental ethic and they didn't want any association with DMC. They were not interested in having anything to do with DMC. . . . DMC operated as 'us and them' whereas Western Nuclear operated as 'we.'

> There's a feeling that [DMC] ripped off the community. There's a lot of anger. That anger is part of the community identity. They are all tired of each other. The benefits were disproportionate. Trust, from DMC not playing by the rules, was lost early on. Sherwood on the other hand, did a lot for the community. They donated trucks and equipment. They heavily involved the community all along the way.

Competing Interests in Millersburg and Albany

Community members depict Millersburg as 'an industrial sanctuary without unduly high regulations and property taxes.' Given the mixture of odors from multiple production facilities, they felt outsiders cast Millersburg as 'that smelly mill town on the freeway' and 'a site of industrial blight, especially before pollution controls were required.' The boundaries between Millersburg and Albany, formally legalized in 1974 when Millersburg became its own incorporated town, permit key informants to make the distinction: 'Millersburg smells, not Albany.' All informants agreed that the paper mill is the largest source of unpleasant odors and most outsiders do not make that distinction. In addition to metals and paper manufacturing, other primary industries in the area include logging, grass seed and crop farming. This creates a 'rural and high tech' mix 'out in the middle of nowhere.' One key informant clearly differentiated Millersburg from Albany such that:

Millersburg is very different from Albany. Millersburg is an industrial area with mainly commercial structures and very few residences. It's an area where farms are being encroached upon and suburbanites are moving out there. Albany is basically a timber town that doesn't have a mill anymore with a lot of low-income people.

While recognizing that the odors are 'a consideration for people thinking about relocating to the area,' one key informant explained, 'when you live here you get used to it, some people call it the smell of money.' Considering that the residential population of Millersburg is only 'about 700,' and its industrial center provides 'about 5,000 jobs' for residents in the surrounding area support for the claim 'they have positioned themselves to develop an industrial base and not much more,' is readily in view.

Despite the persona of being home to successful industrial giants, key informants report that high unemployment rates are no stranger to Millersburg and Albany. Persistent concerns about job losses, including concerns that 'high school graduates are leaving to find jobs,' remains a central focus in community decision-making. A shift 'from resource production to low paying service sector jobs' further complicates matters. Others expressed specific concerns about environmental regulations getting in the way of maintaining jobs. As two community members put it:

> Government needs to cooperate with industry and not drive it away. Let's not have another spotted owl situation in the rare metals industry. We need jobs in the Northwest and additional regulation, especially when regulations are in force but not followed, is an insult to private industry and a disrespect to private enterprise and certainly does not champion our capitalist system. Be a part of our Freedom of Enterprise, not an adversary. Let's work together! The world is moving towards capitalism. Let's set an example! (TWCA Administrative Record 14.4-01026036)

> Every time a business dies as the result of unreasonable regulatory postures, another hand is lost in the effort to clean up and improve the environment. Industry cannot be forced to be environmentally responsible by being forced out of business. When industries die, the public suffers (TWCA Administrative Record 14.4-0011799).

Another community member suggested during a public meeting on September 14, 1993, that even with environmental regulations in place, we can not avoid all risks:

> None of us want to exist in an unsafe environment. We all want safety, reasonable safety. We cannot do away with all the risks in society. When you walk out this door you have no notion that you're going to die – have an automobile accident or something else. No matter what we do in this life, there's risk involved (TWCA public hearing testimony 14 September 1993).

That is not to say Millersburg and Albany residents see no value in environmental regulations, but rather they do not want environmental interests to take precedent over human economic interests. Pollution from their perspective is necessary for progress and can be administratively managed successfully without hurting economic interests or damaging the environment beyond repair. In fact, 'historic neglect of the [Willamette River] is recognized more now' and 'Albany city parks are expanding the recreational use of the river.' Projects to clean up the Willamette River over the last two decades are 'a great source of community pride.' Since local key informants think of Albany and Millersburg as 'a very livable area, close to the mountains, the ocean and Portland,' it is somewhat troublesome to them that 'people from outside don't see the nice areas just one mile away' from the freeway. This leaves decision-makers at a crossroads where retaining the area's strong but 'smelly' industrial base competes with the area's historical American charm, natural beauty and negative images held by outsiders.

Competing Interests in Wellpinit and Ford

The Ford and Wellpinit communities face challenges of balancing industry and traditional, resource-based lifestyles. As one key informant explained, 'they don't want change but want jobs but not large companies.' Not only is 'chronic unemployment in the area' an ongoing issue, but the ore rich terrain leaves many potential doors open for new mining ventures. For some, the Midnite Mine on the Spokane Indian Reservation is 'still a potential, there's a lot of ore left.' Others recognize 'the mine is important to the people that worked there in the past' as it was 'good for the economy but it is 'an environmental liability.' Two key informants suggested, 'there isn't complete agreement about what to do with it' and that 'the tribe has a lot of internal divisions, they haven't spoken with one voice.'

To complicate matters, 'most people won't discuss the fact that there's naturally occurring uranium and radon sources [in the area]. Others accept it as a risk of doing business and living [t]here.' While no one denied that 'mine related chemicals are above background levels,' cleaning them up 'may not reduce them to safe levels, just background levels, just like everywhere else' in the area. As already discussed, the Midnite Mine is not the only uranium extraction and mill operation on the Spokane Indian Reservation. The 'state of the art' Sherwood Uranium Mine and Mill formerly run by Western Nuclear is just five miles from the Midnite Mine. The reclamation of that operation was so successful that the area 'is once again available to tribal members for traditional and historic uses' (Washington Department of Health 2001). The ore rich terrain, although well suited for mining, is not particularly amenable to other development options, given its steep grades. As one key informant stated, 'it's not a place that draws a lot of people for tourism.' Limited job opportunities encourage residents to commute to Spokane for work as well as to supplement personal incomes with food obtained from fishing, hunting, farming and gardening.

Decision-making Time Frames in Millersburg and Albany

The long-standing industrial and economic focus of the Millersburg and Albany communities coincides with short term, immediate personal interests. 'Immediate harm' from environmental regulations that may inhibit local businesses' 'ability to compete at its present levels of employment' is of primary concern. As one key informant put it:

> People are more concerned about themselves. They're concerned if they have milk on the table and if the neighbor's fence is too high, not about hazardous waste issues. . . . Most people aren't in tune with their community. The only people who are in tune are those that seek out specific issues.

The focus of community-level decision-making is on short-term human interests. For example, in letters following a public meeting two community members wrote:

> When the good Lord created this earth, he also created the process of evolution. If your present attitude of preservation remains in effect for the next three or four generations, there will no longer be room on this earth for people (TWCA Administrative Record 14.4-0011771).

> Our community has already been hit hard by the forest products problems. To lose another major industry would be devastating to our economy. . . . In closing, we would like to reiterate the importance of a clean environment, but people must also be considered (TWCA Administrative Record 14.4-0011811).

Economic troubles associated with a recent Teledyne Wah Chang Albany labor strike lasting seven months have 'overshadow[ed] the environmental threats.' One key informant went so far as to say 'the strike was more traumatic than any thing else.' At the same time, key informants raised concerns about temporary replacement workers not knowing how to properly carry out the highly specialized tasks correctly and inadvertently polluting the river. As one explained, 'this work is very specialized. Most jobs can't be learned in a week or two, it takes several years.' The state did in fact, receive more water and air quality complaints than typical during the labor dispute, with a tapering off following its resolution. Generally speaking, one key informant suggested the area residents 'see environmentalists as a culprit in a struggling economy.' Decision-making delays were another source of frustration. One public meeting participant explained that 'this issue has been a controversy for 15 years now. . . .The time has come for a final decision. We cannot afford to wait any longer' (TWCA Administrative Record 14.7-0002).

Decision-making Time Frames in Wellpinit and Ford

The comprehensive decision-making framework of the Spokanes on the other hand, renders concern for both immediate and long-term land use interests. Simply put by one key informant, 'you must take care of the land above all else if it's going to do anything for you.' At the same time, local key informants find it 'hard to understand the technical jargon . . . and the scientific process to know how [industrial activities] affect the cultural process.' With respect to mitigation time frames, all of the key informants expressed frustration with the lengthy processes of identifying environmental risks, posing the question, 'why is it taking so long?' One key informant described the situation as 'going on for years. People get frustrated and become apathetic. It's hard to see what's new.'

Trust in Millersburg and Albany Decision-makers

None of the key informants overlooked TWCA's 'history of past discharge violations' and recognized that 'there still are episodes and occasional releases so it's not erased from memory.' Since 'lots of risks come with the job, like explosions and exposure to chloride gases,' four key informants expressed concerns about worker safety and increased cancer occurrences. Concerns that 'workers don't know what all the risks are' and that 'the company fails to tell them all the hazards of the jobs' contribute to distrust of TWCA. Some workers 'bring in their own bottled water.' During the strike, one key informant suggested 'there have been several minor accidents [during] the strike but they take the injured workers to Salem for treatment to keep it quiet' and that TWCA was 'very closed mouthed about everything.' Two key informants indicated that the 'company fills and caps hot spots and builds on them' so that the spots cannot be tested. When trying to obtain information about company practices, one key informant felt TWCA acted as if everything was 'top secret, they kept no records and didn't tell anyone what they were doing.' This person also suggested that 'the community had a strong sense of the power that TWCA has over their lives' and some workers were 'afraid of repercussions.'

Trust in Wellpinit and Ford Decision-makers

Key informants reported that Wellpinit and Ford residents generally distrust the government. As one put it, 'the higher you go, the worse the smell gets.' One offered that locals are 'paranoid about black helicopters and the government spying on them. . . . They have pictures of all of us and profiles of everyone to develop a strategy to neutralize them.' So far as DMC goes, 'distrust is an issue, there's a long, bitter history with DMC.' Part of the distrust in DMC stems from reports that DMC 'made minimal efforts to work with the community.' Others suggested:

DMC hides under a corporate veil. They have very limited assets and resources, we can't get much out of Dawn. But Newmont has the resources to compensate for the loss of [land] use.

There's not a lot of trust in DMC, they only offered $10 million for the mine clean-up which is estimated to be $160 million.

Some players have been involved in previous law suits against each other. It's been a source of distrust between the state, tribe and DMC.

Distrust in environmental regulators to assure performance criteria are met, fear of radiation in general, and the 'not in my backyard' (NIMBY) attitude typical of many waste disposal issues are 'common reactions' to DMC (Washington Department of Social and Health Services, 1989, p. 1-58). Difficult to understand information can leave impressions of distrust as well. One key informant explained,

I don't really understand the scientific stuff but I trust tribal leadership. I don't have any reason to distrust [other agencies] but don't really understand what they are saying.

At the same time, environmental regulators suggested, 'we want to build trust and engage the community but don't know how.' While trust in federal regulators is minimal at best, trust in local residents involved in making decisions about how to manage the environmental risks associated with DMC is high. One of those local representatives described it this way:

The community doesn't really think about it that much, kind of like 'you take care of it, we don't want to know' and they felt like we were looking out for the welfare of the community.

Summary

The historical identities of the Millersburg and Albany, Oregon communities begin with colonization, individual land ownership, and industrial development. Landmarks of European origins supercede indigenous artifacts both in terms of numbers and importance. Community members uniformly support maintaining the historical role of industry in current community identities, drawing strong ties between industry and community as illustrated in statements like 'the business of TWCA is the business of Albany' (TWCA Administrative Record 14.1-0011833). Community members tend to perceive the pollution associated with Millersburg and Albany industries as both a manageable component of their community identity and a necessary one for economic growth. When outsiders question this, Millersburg and Albany community members are generally quick to defend industry. In fact, several community members suggested that they had more confidence and trust in TWCA than in environmental regulators. This is not true for all community members, however, as some key informants raised concerns about TWCA workers' health. At the top of their list of priorities is securing long-term employment for community members. At the same time, the economic chaos resulting from a recent 7-month labor dispute at TWCA illustrates that they live very much in the present. This focus makes the need for today's environmental regulations and mitigation, frequently seen as excessive, difficult for many community members to fully understand and embrace, especially if mitigation potentially infers increased unemployment. Similarly, frustration with mitigation delays is pervasive and raises concerns about indirect economic and employment impacts.

Central to the identity of the Wellpinit and Ford, Washington community on the other hand, is the maintenance of interactive relationships with the land that precede the arrival of Columbus. Community members tie relationships with the landscape to all activities. Statements like 'creeks are the most important structures' illustrate the use and importance of geographical features in how they define their community. Community is not by any means defined merely in terms of geographical elements, but rather how humans interact with bio-physical landscapes. For example, community members define their community to include Wellpinit and Ford, and frame the DMC Midnite Mine and Ford mill site as one, interconnected site, because that is who and what they interact with. Conversely, environmental regulators distinguish the two areas with geographical boundaries. Incorporated into the community identity of Wellpinit and Ford are nonconsensual changes in land use agreements with the federal government, loss of indigenous hunting and gathering grounds, and in turn, restrictions on the degree to which tribal people can practice traditional lifeways. Not only is the symbolic and moral context of cultural loss a long-standing component of the Wellpinit-Ford identity, but it is also a source of underlying conflict for other issues. Differences in how community members and environmental regulators frame 'the community' may increase the likelihood that symbolic meanings held by community members, but not fully understood by environmental regulators, will contribute to contentious decision-making.

As products of cultural traditions and dependent on the fruits of the landscape for their survival, Wellpinit-Ford community members are protective of natural resources. At the same time, they acknowledge the abundance of mineral rich deposits in the area and view such resources as potential economic assets. The need for development although recognized, competes with cultural traditions, traditional lifestyles and desires for personal privacy. Community members also realize that the landscape in many ways limits their development and mitigation options. They do not oppose all mining activities and have positive experiences with successful restoration of uranium mining and milling operations, e.g. Sherwood Mine and Mill. However, community members also have very negative experiences with uranium milling and mining, e.g., Dawn Mining Company's Midnight Mine and Ford mill site. This makes mitigation a potential vehicle to restore symbolic landmarks and strengthen community identity. At the same time, the life of Wellpinit-Ford community members is one of many competing demands and priorities, potentially making collective responses to risks difficult to formulate.

Preliminary Conclusions

The tasks that lay before today's decision-makers related to the management of potential environmental threats are far from simple. As new information becomes available, we define and redefine what kinds of human-environment interactions pose significant risk to public health and safety.   Simultaneously, communities struggle to balance economic interests, development desires and survival needs. Complicating matters further, ecological and social cycles of recovery from environmental risks frequently operate on different time lines.   How then, can community members, industrial entities and environmental regulators work together to formulate sound, informed decisions that communities can have confidence in?   One step in that process, as suggested by this chapter, is to better understand the role of a community's shared history and identity in community-level decision-making about environmental risks.

On an economic level, Millersburg and Albany, Oregon have a long-standing, generally positive but very dependent relationship with TWCA, a known polluter. This community struggles with the need for mitigation, and it fears that mitigation may result in community-wide economic demise. Conflicting perceptions held by outsiders heighten distrust in regulators and encourage suspicion of mitigation activities.   Being insensitive to such an atmosphere may make mitigation contentious. That is not to say mitigation is inappropriate, but that decision-makers will need to spend time educating the community about why it is important in a familiar language. They will also need to provide ready and timely access to information complied by environmental regulators, experts, and other people participating in assessment and management of the site, about suspected risks and mitigation activities. The community will also need to engage in two-way communication, especially since mitigation is a long-term endeavor.

Ford and Wellpinit, Washington community members have their own set of challenges. They struggle to maintain traditional lifestyles and personal privacy while at the same time recognize a need for economic development.   Some previous courtroom experiences resulted in the Spokane Tribe regaining control over their lands. The tribe is also familiar with the permanent loss of land for non-tribal development.   For them, loss of land and land use restrictions are synonymous with loss of culture. While restoration activities may never return affected lands to pre-mining conditions, mitigation may at least control environmental risks. Only controlling risks, however, is likely to disappoint many. As reported elsewhere, technical experts tend to frame risks in terms of predicting and preventing potential outcomes whereas lay persons focus on detecting and repairing risks. Failing to recognize how different groups frame risks may lead to unresolved conflicts (Elliott, 1988; Gray, 2003; Putnam and Wondolleck, 2003).

Thus, it will be important for environmental regulators to recognize how Wellpinit and Ford community members frame risks and mitigation in terms of symbolic and historical relationships with their landscape in order to develop a dialogue that properly detects and manages potential risks. This will also be necessary to formulate risk communication strategies that are culturally appropriate.

In both the TWCA and DMC cases, mitigation timelines expand across decades. It is not clear how to shorten such timeframes. However, it is clear that slow progress, barely visible on a day-to-day basis, breeds frustration and lack of trust. As stated in chapter one, if distrust in environmental regulators is part of the community identity, community members may be suspicious of the technical or 'official' risk presentations. This in turn, may heighten perceptions of risk, reduce the sense of control community members' have in local decisions, and /or make collective responses difficult to formulate (Frey, 1993; Rosa and Clark, 1999; Williams, 2002). Reversing long-established distrust between communities, polluters, and regulators may not be entirely possible. Perhaps the best that decision-makers can do is to not add more fuel to the fire. One step in that direction may be to make risk and mitigation information complied by environmental regulators, experts, and other people participating in assessment and management of the site, readily and widely available, and in an understandable, culturally appropriate, format. It will be important for the community to determine and communicate what forms and methods of access best suit its members. Likewise, it will be important for environmental regulators to become aware of the community's shared history and identity, including the community's less than favorable past experiences with federal agencies.

Notes

1   The word 'Spokan' originates from the native Salish language, and is generally accepted as meaning 'Sun People' or 'Children of the Sun.' In 1807, David Thompson, a trapper with the Northwest Fur Trading Company, first used the translated name of 'Spokane' when referring to the three Spokane bands (Wynecoop, 1969; Wellpinit School District, 2002). While 'Spokane' continues to be the more commonly used word, many historical documents use the word 'Spokan' when referring to the indigenous people of the Spokane Valley.

2   As stated in chapter 2: The Wah Chang company, a metal alloy producer, took over the United States Bureau of Mines zirconium production in Albany, Oregon, in 1956. In 1967, the Teledyne corporation purchased Wah Chang, renaming the company to Teledyne Wah Chang. In 1972, the company modified its name to Teledyne Wah Chang Albany (Darst et al. 1979). Teledyne Wah Chang Albany (TWCA) is the company's formal name in most Superfund documents, and thus, is the acronym used in this paper. As a company spokesman explained, however, Oremet, another company that produces titanium in southwest Albany, and TWCA were merged when TWCA's parent company, Allegheny Teledyne, purchased Oremet in 1998. While the company name resulting from the merger is now Oremet-Wah Chang, the two facilities remain physically separate with Oremet to the south and Wah Chang to the north. During personal interviews, key informants referred to the company by all four names, with Wah Chang being the most common among local residents.

3   As part of the Superfund process, the Environmental Protection Agency is required to establish an official Administrative Record for each Superfund site. Each item, e.g., reports, letters received, memos, contained in the Administrative Record is assigned a unique number.

Chapter 5

Defining Environmental Risks:
The Case of
Teledyne Wah Chang Albany, Oregon

I am proud of the contribution Teledyne Wah Chang [Albany] has made to the world. I am proud of this wonderful organization and am pleased they have made Albany its home . . . Just who are these [Environmental Protection Agency] people? Anti-progress? Communists? Or just plain stupid? . . . P.S. You probably won't even read this letter but I feel better for having defended a friend of our country (TWCA Administrative Record 14.4-0011828).

Introduction

As the industrial hub of the Willamette Valley for more than 150 years, Millersburg and Albany, Oregon boast of their European origins and diverse economic base. Critical to the foundation of this economic base is Teledyne Wah Chang Albany (TWCA), the area's largest employer. Maintaining the stature of a predominant industrial center requires retaining TWCA and providing conditions under which they can maintain profits and expand markets. As discussed in chapter four, balancing environmental regulations with economic development and a stable employment base are among the top priorities for Millersburg and Albany residents. In efforts to better understand the role of industrial priorities in environmental decision-making, this chapter begins with a description of TWCA's production origins and environmental impacts. It also assesses interpretations of the environmental data available with respect to human and ecological health concerns. Finally, it examines mitigation activities. Lessons learned from this case may provide useful guidelines for managing these challenging wastes as well as related decision-making processes, here and elsewhere.

Teledyne Wah Chang Albany Production Origins

Martin Heinrich Klaproth, a German chemist, discovered zirconium in 1789. It was not until 1824, however, that Jöns Jacob Berzelius, a Swedish chemist, isolated zirconium (Chemical Rubber Company, 2002). Zirconium's high resistance to corrosion by many common acids and alkalis, by sea water, and by other agents makes it an attractive metal for potential use in a variety of industrial processes (Cardin et al., 1986; Chemical Rubber Company, 2002). With an interest in developing widespread use of zirconium and exploring the possibility of developing other metal alloys, the United States Bureau of Mines took over the abandoned Albany College buildings in Albany, Oregon, to establish a branch office and experimental metallurgy laboratory in 1941. At that laboratory, Dr. W.J. Kroll developed a process to produce zirconium from zircon sands in 1947. In 1953 when the Bureau of Mines moved ahead with other metal alloy experiments, the Carborundum Company of America took over the zirconium production activities (Darst et al., 1979).

As the uses of zirconium became better known, demand for the material increased. To meet this demand, the Atomic Energy Commission sought bids for the production of 2,200,000 pounds of zirconium per year for five years, in 1956. Wah Chang, later renamed to Teledyne Wah Chang Albany[1] (TWCA), won the bid and produced its first batch of zirconium sponge on Christmas Day that same year (Mullen, 1971). In addition to zirconium sponge, TWCA also processes the zirconium sponge further to form ingots, tubes, tube hollows, bars, sheets, plates, foil, wire and powder (Darst et al., 1979). When used in conjunction with other manufacturing processes, zirconium products fulfill a variety of unique needs in the energy and weapons industries.

In the nuclear energy industry for example, zirconium metal is a critical component for fuel cladding, the process that results in the production of fuel rods used in nuclear reactors. More specifically, fuel cladding involves filling long thin tubes, normally 12 feet in length, one half-inch in diameter and formed almost exclusively out of zirconium, with uranium-rich fuel pellets (Darst et al., 1979). Zirconium's high resistance to corrosion and low absorption for neutrons makes it an exceptional metal for fuel cladding (Chemical Rubber Company, 2002). In fact, 'zirconium is the only metal used for nuclear fuel cladding in practically all nuclear reactors world wide' (Darst et al., 1979, p. 16). Zirconium has another potential use in the energy production industry. When mixed with zinc, zirconium becomes a magnetic superconductor at very low temperatures. Such superconductive magnets may be capable of producing electric power directly on large-scales (Chemical Rubber Company, 2002). Lastly, in the powder form, zirconium is extremely flammable (Occupational Safety and Health Administration, 2002). This makes zirconium an effective component in armor-piercing ammunition and thus, it is a component in many military incendiary weapons such as cluster bombs (Darst et al., 1979).

In addition to zirconium, TWCA produces other metal alloys including tantalum, tungsten, molybdenium, vanadium and titanium. These, too, are highly resistant to corrosion and heat, making them suitable for use in high-performance jet engines, space vehicles, boats, chemical plant equipment, vacuum tubes, carbide tools, surgical instruments, photo flashbulbs, lamp filaments and even golf clubs. In the power form, they also make effective explosive primers. Filling a unique and specialized worldwide market niche is but one of TWCA's roles. As

Millersburg's and Albany's largest employer, TWCA's presence is vital to the local economy.

Physical Characteristics of TWCA Operations

The TWCA site, along with a particle board plant, a resin plant, a wood flour processing plant, a paper mill, and a closed plywood mill, reside in the heavy industry zone of Millersburg, Oregon. Millersburg was formerly the industrial arm of Albany until its incorporation as an independent city in 1974. The site borders the Willamette River, Truax Creek and Murder Creek. These waterways support recreation activities, fishing, watering of livestock and irrigation. Approximately 30,000 Millersburg and Albany, Oregon residents live within three miles of the site. Hence, with respect to environmental threats and their mitigation, Millersburg and Albany share interests in waste management activities. In reference to other cities and waterways, the TWCA site is about 20 miles south of Salem, 65 miles south of Portland, and 60 miles east of the Pacific Ocean (EPA, 1994b).

Agencies involved in the regulatory oversight of TWCA include the Oregon Department of Environmental Quality and the Environmental Protection Agency and thus, they are the primary sources of information describing the TWCA site in addition to TWCA. Close coordination between these groups yields a common physical description of the site. The areas that comprise the TWCA Superfund site consist of the 110-acre main plant area and the 115-acre Farm Ponds area, approximately 3/4 mile north of the main plant. Sectors targeted for remediation in the main plant area include: 1) the Extraction Area where processing of zircon sands recovered zirconium, one to five percent hafnium, and small amounts of radionuclides such as uranium and thorium; 2) the Fabrication Area where zirconium ingots were transformed into tubes and other products; and 3) the Solids Area where unlined ponds such as Arrowhead Lake, Lower River Solids Pond (LRSP), Schmidt Lake, and V-2 contain waste materials. The Farm Ponds area consists of: 1) four 2.5-acre storage ponds constructed with a soil and bentonite (clay) liner; 2) a 47.8-acre tract that solids from the primary wastewater treatment plant were experimentally applied as a soil amendment in 1975 and 1976; and 3) land used for agricultural purposes.

TWCA's Environmental Legacy

With TWCA's local economic contributions comes a legacy of environmental pollution. The first inspection of the TWCA site completed by the Oregon State Sanitary Authority in June of 1960 'show[ed] all living things killed in Truax Creek for a distance of one mile downstream from Wah Chang' (Darst et al., 1979, p. 2). In May of 1964, inspectors reported 'a massive fish kill at the mouth of Conser Slough, where chemical pollution from Truax Creek hits the Willamette River. . . . and in Murder Creek, which runs through Wah Chang's plant' (Darst et al., 1979, p. 2). Inspectors found another massive fish kill at the mouth of Conser

Slough in April of 1966. One month later, a group of live fish placed in Traux Creek died after being in the water for five minutes (Darst et al., 1979).

In December of 1967, the state of Oregon issued a water pollution permit to TWCA. The permit stipulated a four-mile mixing zone, extending 100 feet beyond the confluence of the Willamette River. This 'inordinately large mixing zone' is the subject of great debate and potentially a contributing factor to fish kills (Kincheloe, 1978, p. 2). In March of 1968, inspectors discovered another fish kill that extended to migrating salmon in the Willamette River several miles away from the plant (Darst et al., 1979). All inspections conducted between 1968 and 1975 revealed violations of TWCA's pollution discharge permit with the first fine, totaling $300, issued in February of 1975. Later the same year, the state of Oregon, Department of Environmental Quality (ODEQ), fined TWCA $5,000 for another 58 separate violations. The ODEQ charged TWCA with additional fines for wastewater discharge and water quality violations in 1978, 1979, 1980 and 1989 (EPA, 1989). All the while, TWCA continued operating, sought more lenient discharge limits, and the four-mile mixing zone remains. Air pollution violations dating back to 1965 when the state of Oregon began to monitor TWCA follow a similar pattern (Darst et al., 1979), with a $4,000 fine for illegal open burning issued to TWCA in 1983 (EPA, 1989). That particular incident involved a test to determine if a waste pile was free of explosive materials whereby, 'an employee tossed a lighted road flare onto a 22,000-cubic-yard pile of hazardous wastes' and it burned for 24 hours. The company justified this action by saying 'safety is our bottom line . . . it would have been negligent for us to have done otherwise . . . we were doing the ultimate check' (The Oregonian, 23 August 1983).

TWCA Superfund Waste Issues

Another long-standing environmental concern at the TWCA facility is the disposal of solid wastes (EPA, 1989). From 1957 through 1979, TWCA disposed processing wastes in four unlined ponds on the facility's grounds: 1) Lower River Solids Pond (LRSP); 2) Schmidt Lake; 3) Arrowhead Lake; and 4) V-2. It was not until 1977, however, that the ODEQ detected significant amounts of radioactive wastes in the ponds, comparable to that of a low-level uranium mill (Oregon State Health Division, 1977a). Information from a former TWCA employee led to the additional discovery of drums containing radioactive wastes buried beneath the sludges in Schmidt Lake (EPA, 1989, 1994b). Since these ponds lie within the Willamette River's 100-year flood plain and are 400 to 900 feet from the Willamette River, their management is of great concern to state and federal regulators (Oregon State Health Division, 1977a; EPA, 1989).

After 1979, the four unlined ponds at the facility's Farm Ponds Area housed processing wastes. In the 1980s, however, changes in production processes reduced the amount of radioactive materials contained in the sludges. While the management of the Farms Ponds is important, public and regulatory agency concerns about the radioactive content of the four ponds located in the flood plain were the focus of the facility's designation as a National Priorities List site under

Superfund. Other issues at the site that contributed to its NPL status include: 1) polychlorinated biphenyl (PCB)-contaminated soil in the Fabrication Area; 2) groundwater beneath the site, particularly in the Extraction and Farm Ponds Areas, characterized as having a very low pH (approximately 1) and elevated concentration of radionuclides (radium, uranium and thorium), metals, volatile organic compounds and PCBs; and 3) surface water sediments within and adjacent to the site contaminated with PCBs, metals including radium, and organic compounds (EPA ,1989, 1994b). While there are several other industries in the immediate area of TWCA, only these wastes generated by TWCA commanded Superfund interest. Said another way, TWCA is the only Superfund site in the area but one among many industries, albeit the largest operation.

Hazard Ranking Scores for Teledyne Wah Chang Albany

When the Oregon State Health Division first detected radioactive and chlorinated residues at the TWCA plant site in 1977, they initially reported that, 'based on staff findings to date, there is no evidence' of any harm to any person from the extent of radiation involved with this operation' (Oregon State Health Division, 1977b, p. 2). Potential leaching into soil and water pathways, however, warranted further investigation. Preliminary assessments of the sludges by the Environmental Protection Agency in 1979 rated the seriousness of the problem as low, on a scale of none, low, medium and high. Assessments completed by Hussein Aldis of Ecology and Environment, Inc in 1980 and 1981 concurred that the wastes at the TWCA site 'do not currently pose any significant threat to public health' (Aldis, 1981, p. 1). Conversely, in 1982 an inspection completed by J. Betz of Ecology and Environment, Inc. upgraded the site's risk to medium. Betz (1982) went on to suggest that the 'land use must be restricted for thousands of years due to radioactive waste' and widespread contamination by radioactive wastes could occur in the event of a small flood (p. 4). Both inspectors noted the numerous failures of TWCA to comply with discharge permit requirements and the issuing of several citations to TWCA from the Oregon Accident Prevention Division. The different inspections conducted in 1982 and 1983 produced a variety of HRS scores ranging from 30.19 to 57.36 on a scale of 0 to 100 (Table 2, Appendix C). As this indicates, HRS scores are subject to debate before a final score is agreed upon for NPL purposes. The TWCA site was proposed for the NPL on 30 December 1982 and formally listed 8 September 1983. While the specific areas included in the TWCA Superfund site are no longer in use, the plant continues to produce metal alloy products and employs over 1,000 workers.

Over TWCA's nearly 20 years of Superfund mitigation, extensive documentation of hazards present on the site are available. Environmental media analyses identified 93 chemicals, including both inorganic and organic substances, on the site, 47 of which pose potential groundwater concerns. Tables C.2 through C.7 in Appendix C summarize the maximum concentrations of specific substances identified on the site (EPA, 1989, 1994b, 1995).

Interpretations of Risks Associated with Teledyne Wah Chang Albany

Potential human health impacts from the volatile and semi-volatile organic compounds identified on the TWCA site involve multiple organs and include increased incidence of dizziness, headaches, fatigue, lack of coordination, vertigo, visual distortion, dermatitis, nausea, vomiting, thyroid insufficiency, liver and kidney damage as well as cancer, and adverse reproductive and fetal development effects (EPA, 1994b). While organic compounds biodegrade, high concentrations may increase the likelihood of long-term effects. When organic compounds are combined, as is the case at the TWCA site, effects may also be more severe, prolonged, and difficult to predict (EPA, 1994b; Klaassen, 1986).

In addition to the volatile and semi-volatile organic compounds, metals and radionuclides are also present at the TWCA site. Unlike organic compounds, metals persistent in the environment infinitely and do not, by definition, biodegrade (Allen, 1999). Their potential to accumulate in multiple sources increases their likelihood of entering the food chain when absorbed by plants in contact with soil and/or water containing them as they migrate between environmental media (Allen, 1999). Furthermore, most metals affect multiple organs including the digestive, cardiovascular and central nervous systems, producing both acute and long-term impacts ranging from abdominal colic to cardiac arrhythmia, renal failure, increased lung cancer mortality and irreversible cognitive deficits (Klaassen et al., 1986). Radionuclides, also present at the TWCA site, irradiate internal organs continuously until eliminated when ingested or inhaled[2]. Organs and systems particulalry affected by radionuclides include the lung, kidney, bone, liver, thyroid, breast and blood cells (in association with leukemia). The extensive half-life[3] of some radionuclides may require their careful management indefinitely (Table C.8, Appendix C). The combination of substances present at the TWCA site makes the potential impacts to human health even more challenging to decipher (Allen, 1999). While it is not explicitly clear what specific adverse short- and long-term human and ecological health effects may result from the substances at the TWCA site, all of the potential health effects involving multiple organs described in this section are applicable.

Individual biological differences influence the severity of health effects experienced at the individual level. In order for contaminants, such as those present at the TWCA site, to produce adverse health effects, one must come in contact with them. This is where predictions become increasingly difficult and highly debated. When assessing the potential health risks from the TWCA site, the EPA considered a variety of different scenarios with various opportunities for exposure ranging from site trespassers, long-term employees, and future residents even though the site is zoned as heavy industry and a change in zoning to residential is highly unlikely. The different low to high risk scenarios depicted an increased risk of cancer in excess of normal background conditions for the area, ranging from zero (no effect) to one in 100,000. By mandate, the EPA considers significant excess overall cancer risk requiring environmental remediation to be one in one million. Since some of the scenarios exceeded the one in one million threshold, the EPA determined that environmental remediation for the TWCA site was necessary. Given the contaminants present, the most probable routes of exposure are inhalation of airborne effluents and ingestion of contaminated foods. As the EPA explained in a public meeting on 6 September 1989:

> The only contact for anybody who's off-site would be if there are dust particles which were emitted off the pond by dust storms, which apparently you don't have down here very frequently. That's the present condition . . .The likelihood of that large of a flood is pretty small. [Then there is] probably the least likely event, but something we have to consider when we're looking at making a decision. This is a condition where we actually have people building homes on the Teledyne Wah Chang Facility next to the ponds which have never been removed. The pond material is dry. So you have exposure through inhalation of dust, ingestion of the pond material and absorption through the skin (EPA public hearing testimony 6 September 1989).

When the scenarios incorporated inhalation, ingestion and dermal exposures, coupled with an unclear understanding of area's climatic conditions, community members attending the public meeting in which EPA presented the scenarios responded less than favorably. For example, following the meeting, community members wrote:

> The use of hypothetical cases such as drinking from an inaccessible source and ingestion of soil containing a low level of contamination do not appear to me to be in the realm of reason. The use of such an approach tests the credibility of the authors, particularly to those of us with a degree of technical background (TWCA Administrative Record 14.4-0011817).

> The EPA assumes risk factors that would exist for a worker 'regularly ingesting soil' at certain areas of the plant site. How many workers at EPA regularly eat dirt? There are **NO** dirt eaters at Teledyne Wah Chang (TWCA Administrative Record 14.1-0011803).

> The severity of the problem has continually been redefined to the point where most of us are confused as to the actual health threat present, if any (TWCA Administrative Record 10.1-0011880).

Recalling another scenario, a key informant reported that 'the risk of building a house on top of the raw sludge and living in it was compared to the risk of traffic accidents from hauling the sludge and the risk from a traffic accident was two to three times greater.' Such comparisons added to the confusion about how the substances present would impact one's personal health.

Since the data about human exposures to the contaminants in question is very limited, one approach to estimating human health risks employed the EPA is extrapolating from animal studies. Two explanations about how they extrapolate human health risks from animal data provided at public meetings include:

And I guess we all know the bottom line, that we have decided there is risk to human health and the environment due to exposure to those contaminants that were identified in the sludge pond . . . So you have to understand that we are extrapolating primarily from animal studies. Although, we do have some studies where people have actually been exposed to contaminants. However, to be protective of people, we believe that it's appropriate to extrapolate from these animal studies (TWCA public hearing testimony 6 September 1989).

There was zirconium found as a contaminant, but the levels need to be very, very high for health effects to occur, so that is not a problem at the site . . . based on laboratory animal studies . . . specific health effects of zirconium are not known (TWCA public hearing testimony 14 September 1993).

One community member attending the meeting responded to the limitations of extrapolating from animal models while excluding the use of other available data in risk assessments with:

It seems unconscionable to me that with the staggering amounts of data generated to date on animal models and theoretical risk exposure that no one has contacted the tumor boards at our three local hospitals as to our actual cancer rates. To my knowledge, there has been no documented increased incidence of cancer in the Linn-Benton area compared to other areas . . . Please let us not chase shadows, but deal with the real problems that face us every day (TWCA Administrative Record 14.4-01026041).

Community members' risk assessment data concerns extended beyond extrapolations. Confusion about the magnitude of the risks also put the EPA at odds with community members as the following dialogue from the 6 September 1989, public meeting illustrates:

Public: Is the material hazardous?

EPA: By what kind of definition? That's an interesting question, and it's a good one . . . The study that was done demonstrates to us that the materials in the sludge ponds are not by regulatory definition hazardous waste material . . . It does contain hazardous substances . . . But by regulatory definition, it is not in total called hazardous waste . . . So does that answer the question?

Public: I would like to know more about the hazardous material we are talking about. Are we talking about the one-tenth of one percent of that that is considered hazardous?

EPA: That's correct. In terms of volume, yes.

Public: A very minute amount.

EPA: Yes. But it does create the kind of risk to—you saw the scale as you go from the average worker on-site to somebody living a lifetime where nothing is done and you build houses on top of the sludge. Sure, that's an extreme case, but that is still a risk to future in habitants (TWCA public hearing testimony 6 September 1989).

In efforts to gain a further understanding of the risks present in a public meeting on 14 September 1993, a community member asked:

Can you put the levels of radiation in terms we can understand? I mean, compare it to something the we would have an idea of, if you're talking about for instance, a lantern mantle or brick wall or something we are familiar with (TWCA public hearing testimony 14 September 1993).

EPA responded with another restatement of the risk assessment and exposure routes. The response followed by this explanation was:

Public: So you're saying you didn't compare it to anything. Is it like a chest x-ray once a year or once a day or once a month, or carrying a lantern mantle around in your pocket? What is it like?

EPA: I don't have a comparison like that but what I do have is the comparison to background. Everything was compared to background, so it's higher than what would be a normal background level for this area of the country, for Corvallis actually. The radiation levels in that screening were taken from the Corvallis Airport, so it's higher than that.

Public: Sure, but so is the level of soot in the air from filed burning.

EPA: That doesn't pose a radiation risk.

Public: I don't understand what the risk associated – you're not – you're still not answering the question. You're not comparing it to something we know about.

EPA: Okay. Well, then why – that could be a comment that I could provide very specific information for you if you'd like to write that down and give me someplace specifically to answer that back to you, I can provide that information (TWCA public hearing testimony 14 September 1993).

Hence, the public's question about personal health risk remained unanswered, heightening rather than resolving community members' frustration with confusing risk information. Leaving questions unanswered may also increase distrust in experts and federal officials (Frey, 1993).

There was also a lot of confusion over what an increased risk of one in one million meant. After two failed attempts to explain it, EPA offered the following analogy to the audience:

If an exposure occurs, then everyone who has an exposure, you could think of them having a little grid laid over them, like I'm a little cut-out doll, you lay a grid and it has a thousand – a million little squares in that grid. You color in one of those little squares in that grid, and that's my risk. Out of all – I have a million chances. I have a grid laid over me that has a million little squares in it, and one of those squares is colored in. My risk is one in a million, that's my probability of getting cancer. Every single person has that risk. It's not like one person out of a group of one million, or 10,000 or 1,000 people is going to have cancer. Everybody has the same probability (TWCA public hearing testimony 14 September 1993).

After three more attempts to explain it in a way the audience could understand, the EPA moved on to another question, leaving the matter unresolved. Later in the discussion, the TWCA environmental lawyer revisited the issue, explaining it in way that the audience was able to understand. This resolved the issue for EPA in part, as well as created an opportunity to build upon his own agenda. His explanation was:

The one in a million that we've talked about before, the one dot out of a million on you, well, I guess, using that example, you've already got 250,000 dots colored in right now because that's your chance of getting cancer. The one more dot of drinking water that you can't get to, or eating soil which you're not going to eat, well, you draw your own conclusion (TWCA public hearing testimony 14 September 1993).

Emphasizing the lack of clarity in EPA's risk communication efforts, one of the final comments at the meeting regarding explanations of human health risk was:

I believe your presentation today fails the public in communicating the risks, and therefore there's no ability on our part to make any sense of it. And from what I read in various articles in several journals, one of your dominant responsibilities is to make sure that you're credible with the public, and you haven't done that, because you haven't talked about what kind of exposures and actual risks or averages (TWCA public hearing testimony 14 September 1993).

As this illustrates, incomplete and unclear answers to the public's questions encouraged distrust in the EPA. The lack of agreement about the riskiness of the substances present also brings the need for mitigation into question, especially if mitigation infers job losses, an essential underlying issue for Millersburg and Albany residents.

Another topic of debate at the same public meeting was the failure to recognize the remediation activities TWCA already completed. More specifically, the environmental legal council for TWCA criticized EPA for not recognizing how much TWCA had done already and commended TWCA for removing the sludge in 'one-third of the time that EPA had ordered it be done in' (TWCA public hearing testimony 14 September 1993). He also questioned how the risk assessment completed by the EPA focussed on past, rather than current practices, and pointed out that 'this is a manufacturing plant with people working in it constantly all day long. This is not Love Canal, this is not a hazardous waste site in the middle of nowhere that's been abandoned' (TWCA public hearing testimony 14 September 1993). He expanded upon this idea by saying:

> There is complete recognition of the fact that after 50 years' worth of operation, chemicals have been spilled, metals have been cut and put aside, things have been buried. It is a plant. And, God willing, it will continue to be a plant. The focus of this study was not to say, what do we have to do now to [TWCA] to turn it into a residential housing development. There's no plan for that, and that was not the purpose of the study . . . Those are not pumping wells. [TWCA] is not a drinking water company. [TWCA's] employees use water from the city . . . What's the risk of that soil anyway? The risk of that soil is only a problem if somebody is exposed to it. Well, how are you exposed to that soil? Well, in their 15-page plan, they say they're worried about workers perhaps eating sandwiches with dirty hands, and eating some of that soil. They're worried about workers smoking a cigarette with dirty hands, and eating some of that soil. Well, let's think about that for a moment, folks. [TWCA] provides lunchrooms for its employees. Those lunchrooms are clean places; they're sanitary places. They have facilities to wash their hands. I think the risk of worrying about workers eating soil, ingesting the soil, is almost an insult to the employees of [TWCA] and to its management. If it's not an insult it's certainly an unnecessary act that we need to dig up all that material and move it away (TWCA public hearing testimony 14 September 1993).

In conclusion, he encouraged TWCA employees at the meeting to embrace this position by stating:

> Send in comments and let EPA know how you feel about this proposed plan here. After all, this is your community. You live here, many of you work at the plant, you should take control of the situation. Finally, I think we'd like to get this back on a proper track. Frankly, I'd like to see the lawyers go home. Good night (TWCA public hearing testimony 14 September 1993).

As these dialogues suggest, TWCA's status as the area's largest employer harnessed support from community members as well as put TWCA in a largely unquestioned position to dominate the risk decision-making activities. Unclear communication of risks on EPA's part and worries about job losses reinforced support for TWCA's position.

Another aspect of risk communication that Millersburg and Albany residents responded to was the position of the messenger in relationship to the community, i.e. an insider or an outsider. Community members expressed more trust in TWCA than outside environmental regulatory agencies as well as frustration with EPA's lack of recognition for TWCA's long standing community contributions, particularly in the form of employment. For example, in letters following the 14 September 1993 public meeting, a community member wrote:

TWCA has been a part of my life since its inception in 1956. I believe Teledyne is doing everything possible to ensure the safety and health of this community. They have been concerned with quality of life long before the Environmental Protection Agency imposed any restrictions. They continue to act responsibly and responded to community desires and needs (TWCA Administrative Record 14.4-0011822).

Drawing attention further away from TWCA's past practices, another community member wrote:

From my experience, including four years in the U.S. military, the worst polluter in the United States would appear to be the United States Government. Private industry has taken the lead in volunteering to clean up its own facilities. The last thing we taxpayers need is any further waste of our tax dollars by what appears to be absolute overkill . . . I feel having observed Wah Chang for 20 years that they are a good neighbor and worthy of my trust (TWCA Administrative Record 14.4-0011788).

Six key informants supported this perspective by describing TWCA's waste issues as 'all in the past and taken care of now.'

Seven other key informants, however, suggested that workers at TWCA have a different perspective and raised concerns about increased cancer rates (particularly prostate and mesonephroma, a relatively rare tumor involving the mesonephric cells in reproductive organs) and asbestosis. Key informants also reported that workers recognize 'lots of risks come with the job, like explosions, exposure to chloride gases' but workers were concerned that the 'company fails to tell them all the hazards of the jobs' and as a result, 'workers don't know what all the risks are.' Many workers 'bring in their own bottled water' as the result of workplace safety concerns. In some ways, health concerns promoted a new form of cohesion among workers. More specifically, one key informant reported that 'people came together with the strike and about health concerns, it's not just about accidents anymore.'

Fish kills are not necessarily a thing of the past either. One key informant recalled a 'big fish kill three years ago.' Another key informant suggested, 'it's always in the back of your mind, there's a fear of an accident . . . I don't live there [Millersburg], I don't know that I would feel safe there.' On the other hand, one key informant was 'more concerned about alcohol and drug problems than this.' One letter submitted following a public meeting stated, 'this is not Love Canal by any stretch of the imagination.' While the HRS scored for Love Canal and TWCA are similar (54 and 54.27, respectively), many people were relocated at Love Canal whereas TWCA continued to operate much in synchrony with normal routines carried out for decades. No noticeable disruption in daily activities was evident in Millersburg and Albany nor did key informants report that people were leaving the area in response to potential health concerns. At the same time, some people wonder, as one key informant put it, 'What are the long-term effects? I don't know, nobody knows.' Others argue the 'perception about radiation is over-represented as dangerous.' Similarly, one key informant suggested 'some people are quick to jump to the conclusion that "there's an odor, I'm at risk for cancer" and that's generally not the case.'

Mitigation Actions for Teledyne Wah Chang Albany

Mitigation activities at TWCA are near completion. Ongoing activities involve the treatment of groundwater to reduce the presence of contaminants to a cancer risk of 10^{-4} and to meet discharge permit stipulations. Completed activities include: 1) the removal of sludges from Schmidt Lake (15,000 cubic yards), the Lower River Solids Pond (75,000 cubic yards), and V-2 (4,500 cubic yards) to an off-site facility; 2) removal of 3,600 cubit yards of contaminated sediments in surface waterways; 3) slope erosion control involving the placement of riprap on the banks of Traux Creek to prevent contaminated materials from entering the creek; 4) the institution of formal deed restrictions such that the groundwater and land will be used only for industrial purposes; and 5) the institution of long-term surface and groundwater monitoring (EPA, 1989, 1994b). In addition, the removal of 1,869,192 square feet of contaminated soil in the sand unloading station in the Fabrication Area, front parking lot in the Extraction Area, Schmidt Lake in the Solids Area and the entire Soil Amendment Area was necessary to reduce gamma emitting materials to 20 microrem/hour above background (or 4pCi/liter). These are potential building sites for industrial activities. The clean-up level of 4pCi/liter meets indoor radon concentrations acceptable for industrial buildings. Hence, the purpose of this standard is to minimize potential future problems. Finally, in 1999 TWCA purchased four residential properties potentially affected by the site but no other residential relocations are under consideration (EPA, 1995, 2001b).

The cost for Teledyne Wah Chang Albany (TWCA) to complete these actions is roughly $20.5 million to date, with another $5 to $10 million to go for the groundwater treatment (EPA, 1989, 1994b, 1995, 2001b). The selection of these actions over less expensive alternatives, ranging from no action to containing the wastes in place, has been the subject of great debate. One of the grounds for the debate stems from an unclear understanding of what the potential risks to human health from the wastes actually are as the previous section illustrates. Given the complexity of the site, even experts find the 'rigorous chemical processes difficult to understand unless you have a Ph.D. in chemical engineering.' This is complicated by the fact that in the early years of operation, 'the standard disposal practice was to pour [wastes] into gravel areas, even federal regulatory agencies encouraged backyard disposal at that time.' The 'mountains of paperwork', 'expensive, long, drawn out, ongoing saga,' and 'lack of trust among the parties involved' add to the frustration key informants have with the mitigation's seemingly slow progress and delays. One key informant took this a step further and suggested 'public interest in this site has been from environmental groups supporting the clean up and the TWCA employees wanted to leave things alone.'

Prior to one public meeting held by the EPA to describe the clean up alternatives for Schmidt Lake and the Lower River Solids Pond, a key informant reported that:

TWCA hired a [public relations] firm, wrote a 'white paper' outlining why the EPA was wrong and could end up shutting down the plant, mailed it to all of its employees and other key community leaders, and then orchestrated public comment against the EPA cleanup.

The report focused on TWCA's long standing contributions to the community and detailed all of the remediation activities TWCA completed to date, in spite of the fact, as TWCA argued, adverse human health effects associated with the substances present were still unclear. In the report, TWCA raised concerns about being able to retain the plant in Millersburg, given the increased use restrictions and the negative financial impacts to the company from, in their opinion, excessive remediation actions imposed by the EPA. Finally, TWCA raised concerns that the remediation plan selected by the EPA and presented to the public for comment, alternative 7, was not included in the draft public documents. Even though alternative 7 was included in the final proposed plan, legal council for TWCA drew additional attention to exclusion of alternative 7 in the draft documents, posing this question at the 14 September 1993 public meeting: 'why is it that after six years, after two versions of that massive study over there [he said over 99,000 analyses had been done earlier in his statement], is there no alternative seven [the alternative EPA selected], the seven that pops up in this 15-page paper?' Representatives of TWCA also strongly opposed the five-minute comment limit at the public meeting. The purpose of the time limit from the EPA's perspective was to give everyone who wanted to comment on the plan an equal opportunity to do so. However, this inspired community members to offer statements following the 14 September 1993, public meeting in support of TWCA's perceived excessive remediation requirements and right for additional time like:

> I believe the EPA is not being realistic about the clean-up. It took many years to get where we are now and it will take many years to undo the problems. It cannot be done overnight. [TWCA] has been trying to meet all the requirements–why not allow a reasonable length of time to clean up, with progress measured periodically? Please keep in mind, [TWCA] is a major part of the community and one of the largest employers in Linn County (TWCA Administrative Record 14.4-0011849).

> As an employee of the EPA why risk public humiliation for failure when you can set the standards so high the chance of a clean up failure is almost nonexistent. The rationale is very good if you are protecting your job, but is it rational for the thousands of people who may loose their jobs just to save a few . . . Most people would agree a perfectly clean environment would be wonderful, but what would the cost be? People living in only warm areas and grass huts and wearing 100% nature made clothing? I feel most people in the area of the plant would be willing to live with the slightly increased risk of death while cleaning up the area in a manner that is compatible with [TWCA] continuing operations (TWCA Administrative Record 14.4-0011782).

Making this company spend millions of dollars chasing after minuscule risk factors based on unrealistic exposure assumptions is intolerable regulatory policy . . .This is like prescribing an expensive medicine for a headache which an aspirin would cure. And why did you refuse to give TWCA the courtesy of adequate time to state their case at the public hearing??? Only 5 minutes, I can't believe it. I've followed the regulatory process of your agency and the cooperativeness of the plant for many years. I just can't understand why you are squelching reasonable input from the involved party (TWCA Administrative Record 14.4-0011837).

Preliminary Conclusions

As we learned in chapter four, the Millersburg and Albany, Oregon, community is very dependent on TWCA in order to maintain its long standing reputation as Willamette Valley's industrial hub. Key informants reported that the majority of Millersburg and Albany residents work and live in the immediate area. This provides the opportunity to develop cohesive relationships between community members. In fact, key informants reported that over 60% of TWCA employees have been with the company for more than 25 years. Inadvertently, this may reinforce uniform fears that too much mitigation may result in community demise. While key informants suggest some workers have concerns about occupationally related health risks, 'nobody wants to close the company down' just the same.

The role of TWCA as the area's largest employer is not only important to the local economy, but also provides TWCA with the opportunity to dominate community-level decision-making. Community members appear to readily support such a position, sometimes denouncing the input of outsiders, particularly the EPA. Moreover, the clearly defined lead authoritative roles and top-down management strategies of outside agencies offer community members limited opportunities to influence decision-making, making it a process of informing, rather than actively involving community members, from their perspective. Alienation of community members in the decision-making process increases when EPA experts leave many of the community members' questions, albeit questions entrenched in uncertainty, unanswered. As one community member wrote following the 14 September 1993 public meeting:

I was very disappointed that your staff did not answer the Public's Questions more directly. Did they not know? Were they trying to PULL THE WOOL OVER THE PUBLIC'S EYES? Were they trying to make a job for themselves? (TWCA Administrative Record 14.4-0011851).

Such circumstances create a sense that outside expert knowledge offers little to the solution and that outside experts may be withholding information, building upon established relationships of distrust. An incomplete understanding of the area's geographical conditions on the part of the EPA further reduces confidence and trust in experts' assessments. Under such conditions, establishing cohesive relationships between community members and environmental regulators is very challenging at best, and gridlock is very possible (Rosa and Clark, 1999; Williams, 2002).

At the same time, these factors may make it difficult for experts to dominate the policy process. The little media attention given to TWCA issues may have unexpectedly limited the opportunity for some groups to advance their position over others and helped keep everyone's interests on the same playing field. On the other hand, the media can 'produce a false consciousness that legitimizes the position and interests of those who own and control the media' (Anderson, 1997, p. 21). Thus, little media attention may reflect efforts to down play or silence concerns about TWCA and indirect support for TWCA by those who control the media. Regardless, EPA's federally mandated lead authority role in decision-making, while perhaps not deemed legitimate by some community members, allows the EPA to remain in control. In turn, lack of trust in environmental regulators appears to heighten fearfulness of long-term risks from the TWCA site, as well as reinforce a sense of little influence in decision-making among community members.

The contentious decision-making atmosphere is complicated by the fact that the health effects associated with the TWCA site are not clear. Data about specific health effects from the mixture of contaminants present is not available. Extrapolations based on animal studies and estimated probabilities rather than known outcomes provided guidance for risk decisions. While such risk estimation practices are common in technical circles, they are frequently confusing to lay persons, especially when communicated in unfamiliar terms and accompanied with complex analogies, as was the case here. The information utilized in the TWCA case also included some scenarios community members deem unlikely, e.g., residing on the Schmidt Lake site, potentially adding to the confusion. Such circumstances may increase fear of potential hazards for some. Others argued cause for concern is not necessary as administrative controls limit access to the substances present. In addition, several other industries are present in the immediate area so it is impossible to eliminate risk completely. Furthermore, as one key informant put it, 'the average human being doesn't concern themselves with risk, that's why we drive too fast and drink too much.'

On the other hand, those that expressed concerns about increased cancer in relationship to TWCA employment, interpret risk reduction largely as a personal responsibility that comes with the job and engage in individual responses, e.g., bringing bottled water to work, rather than collective responses. However, one of the issues under negotiation in the recent strike was maintaining health care benefits for retired workers, three of whom were recently diagnosed with cancer according to key informants. Key informants suggested that the strike encouraged workers to discuss long-standing health concerns related to TWCA employment previously managed on an individual basis. To that end, the strike provided an opportunity to develop cohesive relationships within subgroups formed as a result of common and very specific interests. The close physical proximity and utilization of common locations offer support to maintain such cohesive relationships. While there is no clear mass exodus from the area as a result of health or ecological concerns related to the TWCA site, one should not infer all community members have become at ease with the risks present just the same.

In this Superfund case, TWCA is clearly held accountable and accepts responsibility for the hazards present. Mitigation activities thus far expand across two decades. However, the long-term management of the risks present does not come with a guaranteed end date. Given the types of contaminants involved, the potential for a failure in current containment measures and/or administrative controls, and the potential discovery of new problems leaves the mitigation door open indefinitely. The ability of everyone to remember the location of the hazards as new industrial development takes place poses another challenge, further complicated by the little confidence local people have in environmental regulators.

What we can learn from this case, however, is that creating opportunities where communities maintain some sense of influence in mitigation decisions, largely made by external agencies, may help minimize conflict during decision-making processes. This is especially important in situations where risk outcomes are uncertain, frequently making the need for and type of mitigation that is necessary difficult to determine. Under such circumstances, the authority of outside entities to make local decisions is likely to be scrutinized (Putnam and Wondolleck, 2003), as was the case here. Moreover, mitigation actions that may seem unnecessary or difficult to understand may threaten established community identities that community members do not desire to change e.g., TWCA, the area's largest source of employment. Threatening the identity of a group may call into question the beliefs and values of its members (Gray, 2003). In turn, debates about risks may unintentionally expand into debates about group legitimacy and encourage defensive behavior. It is also important for external agencies to be cognizant of local power structures. Such structures may impede decision-making in unexpected ways if not understood and respected, e.g. the white paper TWCA sent out to their strongly supportive workforce prior to a public meeting. That is not to suggest that external agencies should cater to community desires without question or vice versa, but rather each needs to recognize the position of the other with open ears.

Another important consideration is developing strategies to communicate uncertainty in ways that community members can understand, allowing them to make better informed personal and collective decisions. This is especially important given the enduring nature of the hazards present in the Millersburg and Albany, Oregon landscape as the uncertainty associated with such substances will most likely remain unresolved for many years to come. That is not to suggest that decision-makers and community members should simply look beyond their differences and move on, but rather they need to recognize their differences and learn as much as possible from them in order to move ahead. At the same time, one must develop realistic expectations about erasing the ingrained distrust among those involved in the decision-making process and reversing top-down risk management approaches overnight, if at all. These are not simple issues; hence, simple solutions are not likely to be successful. As one public meeting attendee proposed, 'we need to be adult about this and do what is in the best interest of everyone involved and not let our immediate needs compromise this responsibility' (TWCA Administrative Record 14.4-0011814).   At the same time, we must recognize the increasing complexities and competing demands that communities struggle with, especially under conditions of growing economic and environmental risk uncertainty.   This may be even more challenging for some communities than others. Perhaps as one community member advises,

> In this competitive world we must not only protect and preserve our environment; we must also protect and preserve our ability to compete and provide employment to our citizens. We must assure a balance that equally addresses all of these concerns (TWCA Administrative Record 14.4-01025043).

Notes

1   As stated in chapter 2: The Wah Chang company, a metal alloy producer, took over the United States Bureau of Mines zirconium production in Albany, Oregon, in 1956. In 1967, the Teledyne corporation purchased Wah Chang, renaming the company to Teledyne Wah Chang.   In 1972, the company modified its name to Teledyne Wah Chang Albany (Darst et al. 1979). Teledyne Wah Chang Albany (TWCA) is the company's formal name in most Superfund documents, and thus, is the acronym used in this project. As a company spokesman explained, however, Oremet, another company that produces titanium in southwest Albany, and TWCA were merged when TWCA's parent company, Allegheny Teledyne, purchased Oremet in 1998. While the company name resulting from the merger is now Oremet-Wah Chang, the two facilities remain physically separate with Oremet to the south and Wah Chang to the north.   During personal interviews, key informants referred to the company by all four names, with Wah Chang being the most common among local residents.

2   Ionizing radiation displaces electrons, causing the element to be chemically active. There are three kinds of ionizing radiation, alpha, beta and gamma. Gamma radiation affects all organs and is the most penetrating but can be easily deflected and quickly decreases with increasing distance. While alpha is less penetrating than beta, if inhaled or ingested, alpha and beta radiation irradiate internal organs continuously until eliminated. Routes of exposure that are of greatest concern are inhalation of airborne effluents and ingestion of contaminated foods.
40 Code of Federal Regulations (CFR) 190 environmental radiation protection standards for nuclear power operators:
     75 millirems per year for the thyroid
     25 millirems per year for whole body and other organs
     (Only applies to uranium 238 and 234, radium 226, thorium 230)

3   half-life: time in which half of the atoms of a radionuclide disintegrate into another nuclear form; also the time required for the body to eliminate half of the material taken in.

Chapter 6

Defining Environmental Risks: The Case of the Dawn Mining Company, Washington

Radionuclides are inherently scary. They have an image of it being bad stuff in our community and people wonder, are we being poisoned? People that live off the environment are concerned about meat and the water supply but there's a lot of undefined anxieties. I don't have to live there so it's hard to understand but I try. People ask how safe is it and say I don't want it in my backyard (Key Informant).

Introduction

How does a uranium mine end up amidst a culture with close ecological and spiritual connections to the land like an Indian reservation? As discussed in chapter four, part of the reason stems from the fact that two-thirds of the uranium ore deposits in the United States reside on Native American lands (Kuletz, 1998; Schulz, 2001). For the Spokane Tribe of Indians, uranium is one of 24 metallic minerals and harvestable economic assets found in the harsh terrain of the aboriginal lands they managed to retain (Stevens County Rural Development Planning Council, 1961; Wynecoop, 1969). Under the plague of chronic unemployment, limited development options, and only a cursory knowledge of hazards associated with uranium extraction coupled with highly favorable market conditions in the 1950s, the more relevant question became, why not? When ore prices fell and knowledge of hazards increased over time, acceptance of this industry as a viable economic endeavor was no longer clear. In efforts to better understand the transformation of risk perceptions and development of new conflicts among environmental regulators, industrial entities and community members, this chapter describes how uranium mining began and ended. It also assesses interpretations of the environmental risk data available. Finally, it explores the mitigation planning challenges that lie ahead. Lessons learned from this case examination may identify useful guidelines that can assist with the management of such challenging issues here and elsewhere.

Dawn Mining Company Production Origins

In 1942, the United States committed itself to an atomic weapons program as a means to end World War II (Williams, 2002). This effort, more commonly known as the Manhattan Project, produced the first atomic weapons. Detonation devices developed at national laboratories in Los Alamos, New Mexico, used in conjunction with enriched uranium from Oak Ridge, Tennessee, and plutonium from the Hanford Nuclear Reservation north of Richland, Washington, formed the bombs dropped on Hiroshima and Nagasaki, Japan in August of 1945 (Dalton et al., 1999). Commitment to atomic weapons development did not end with Japan's surrender but rather expanded during the Cold War that followed. In order to support such efforts, the Atomic Energy Commission, founded in 1946, developed uranium procurement and exploration programs with attractive government incentives and guaranteed ore prices for would-be investors. By 1954, more than 500 uranium mines were in operation in the Colorado Plateau alone. Some refer to this era as the greatest 'get-rich-quick' mining boom the United States ever experienced (Hahne, 1989).

With such promises of fortune, uranium ore exploration quickly extended beyond the Colorado Plateau. Two hopeful prospectors, the Spokane tribal members and brothers Jim and John LaBret, discovered a fluorescent material near Spokane Mountain on the Spokane Indian Reservation in the state of Washington in 1954. The Bureau of Mines identified the substance as meta-autunite, a uranium rich mineral. Shortly thereafter, the brothers founded the company, Midnite Mines, Inc., to further investigate their discovery. Exploratory drilling confirmed the presence of extractable ore quantities. In need of a financial backer to extract and mill the ore, a partnership between Newmont Mining Corporation (51% ownership) and Midnite Mines Inc. (49% ownership) formed the Dawn Mining Company (DMC). Ore extraction began in 1955 following the establishment of a mineral lease agreement through the Bureau of Indian Affairs that granted DMC access to 811 acres of land centrally located on the Spokane Indian Reservation, near Spokane Mountain (Ruby and Brown, 1970; Wynecoop, 1969). The area specified in the lease consists of land held in trust by the Spokane Tribe of Indians, including the Boyd Family allotment (DeGuire, 1985; Shepherd Miller, Inc., 1991). In 1956, DMC completed the construction of a mill, located approximately 18 miles away from the mine. The mill site occupies part of an 820-acre plot DMC owns in Stevens County near Ford, Washington, adjacent to the reservation's east boundary, but entirely off the reservation (Washington Department of Social and Health Services, 1989).

From 1956 to 1965, DMC held a contract with the U.S. Atomic Energy Commission to produce uranium oxide (U_3O_8), or yellow cake, an important feed used in uranium fuel enrichment plants and fuel pellet fabrication. Two unlined, above-grade, tailings disposal areas, TDA 1 and TDA 2, contain approximately 1.2 million tons of tailings wastes generated from these contracts and occupy 59 acres near the mill site. At that time, such disposal practices were acceptable and provided a means to remove settable solids from drainages as well as recover water for reuse at the mill (Washington Department of Social and Health Services, 1989). The DMC's Ford mill site processed ore extracted from other mines, including the Silver Buckle Mine (later known as the Sherwood Mine and Mill) near the Midnite Mine, under contracts with mine owners. Tailings resulting from these private contracts reside in TDA 1 and TDA 2 as well. In 1965, decreased demand and dropping uranium ore prices led to the termination mining and milling operations (Washington State Department of Health, 1991).

Following an improvement in the uranium market and a one million-dollar renovation of the mill in 1969, DMC produced yellow cake under commercial contracts with electrical utility companies until November of 1982. Tailings generated between 1969 and 1981 were disposed in a third unlined, above-grade impoundment area, TDA 3. This 46-acre impoundment area is adjacent to TDA 1 and TDA 2. In 1978, however, the integrity of the safety berm around TDA 3 diminished, raising concerns about overflow and the possibility of heavy rains cutting a channel through the dike. As a result, the construction of a fourth impoundment area capable of accommodating tailings generated from the remaining 973,000 tons of proven ore reserves with a uranium dioxide grade of 0.143% at Midnite Mine, began in 1979. In contrast to the other impoundment areas, this 28-acre tailings disposal area, or TDA 4, is adjacent to TDA 3 and 65 feet below-grade with a liner as specified by regulations of that time (Washington Department of Social and Health Services, 1980). Upon its completion in 1981, TDA 4 housed all new wastes generated by milling operations (Washington Department of Social and Health Services, 1989; Washington State Department of Health, 1991). Experts believed, the uranium oxide produced as a result of the continued mill operation would 'significantly [contribute] to the nation's electrical energy production in a time of pointed attempts to achieve national self-sufficiency in energy resources' (Washington Department of Social and Health Services, 1980, p. 10-9). Since the percent of electricity generated from nuclear fuels was projected to increase 'from 8.6% in 1975 to 26% in 1985', support to expand the tailings ponds was abundant (Washington Department of Social and Health Services, 1980, p. 6-6).

Later that same year, the Spokane Tribe of Indians suspended mining activities after discovering DMC violated the extraction specifications of the lease, only removing selective high-grade ore. The extraction strategy employed by DMC at that time also raised concerns about excessive erosion in some mine site areas (Washington State Department of Health, 1991). In 1982, an unexpected drop in uranium prices, coupled with the suspension of ore extraction at the Midnite Mine, led to a termination of milling activities, approximately six to ten years sooner than DMC planned. Ore prices remained below extraction costs, and mining and milling activities never resumed. In its years of operation, the DMC recovered approximately 11 million pounds of uranium dioxide (U_3O_8) at its mill site located near Ford, Washington from 2.9 million tons of ore extracted from Midnite Mine (DeGuire, 1985; Shepherd Miller Inc., 1991). The grades of ore milled ranged from 0.10% to 0.25% with an average of 0.225% (Washington State Department of Health, 1991).

Physical Characteristics of DMC Operations

The area developed between 1955 and 1981 for the Midnite Mine is approximately one-half mile wide and one mile long, and covers about 321 acres, centrally located on the Spokane Indian Reservation (Marcy, 1993; Marcy et al., 1994; US Bureau of Mines, 1994). Two pits remain open. Seep waters from the mine, containing radionuclides and heavy metals, feed three drainages on the site that lead to Blue Creek. Blue Creek flows into the Spokane River arm of Lake Roosevelt, a popular fishing and recreation area. According to the Spokane Tribe, 1,230 tribal members live on the Spokane Indian Reservation, almost all of whom live within a 17-mile radius of the site. Moreover, 140 tribal members live along the mine's hauling route (EPA, 2000d).

The EPA is currently the lead agency managing the Midnite Mine site. Prior to the EPA's involvement, Bureau of Mines and the Bureau of Indian Affairs managed the mine site. When Congress eliminated the Bureau of Mines as an agency in 1995, the Bureau of Land Management (BLM) assumed their responsibilities. As a result, data from a variety of investigations completed by multiple entities are available for the Midnite Mine and descriptions about the mine vary somewhat among them. For example, annual precipitation at the mine site, an especially important consideration for the management of open pits, range from 17.3 to 20 inches (Table D.1, Appendix D). Reports indicate the Midnite Mine is 35 to 50 miles from Spokane, and 5 to 8 miles from Wellpinit, the closet community to the mine and home to most of the Spokane Tribe of Indians headquarter offices. The distance of the ore hauling route between the mine and the mill is 18 to 25 miles (Table D.2, Appendix D). Similarly, the distance from the mine site to Lake Roosevelt, an upper portion of Spokane River, range from 3.5 miles (BLM, 1996, p. 2528), to 4 miles (U.S. Bureau of Mines, 1994, p. 2), 4.5 miles (Superfund Technical Assessment and Response Team, 1998, p. 2-2), and 4.6 miles (URS Corporation , 2000, pp. 3-13, 3-14). The elevation of the mine site ranges between 2,400 and 3,570 feet (Table D.3, Appendix D). Descriptions of the ore body at the mine site also vary from 55 to 600 feet wide, 700 to 1,200 feet long, 140 to 150 feet thick, and 15 to 300 feet below the ground (Table D.4, Appendix D).

There are several areas where descriptions of the physical parameters of the Midnite Mine are consistent. For example, reports uniformly reveal that the area has no history of serious flooding or seismic activity (Shepherd Miller, 1991; Washington State Department of Health, 1991; URS Corporation, 2000). Most importantly, agreement exists about the specifics of the 321 acres developed for mining. Approximately 33 million tons of waste rock and 2.4 million tons of low-grade ore containing an estimated two million pounds of uranium oxide (U_3O_8) remain stock piled on 84 acres at the Midnite Mine site (Shepherd Miller Inc., 1991; U.S. Bureau of Mines, 1994; Superfund Technical Assessment and Response Team, 1998). Two pits, Pits 3 and 4, remain open. Pit 3 is roughly 550 feet deep with a surface area between 9 and 10 acres. Pit 4 is about 450 feet deep with a surface area between 5.5 and 6 acres (Superfund Technical Assessment and

Response Team, 1998; URS Corporation, 2000). Other primary sources of potential environmental contamination on the site include the Pollution Control Pond and the Blood Pool. The Pollution Control Pond is a man-made catchment basin for mine seeps with a surface area of 0.9 acres, a depth of 30 feet, and holding capacity of 2.5 millions gallons of water (Sumioka, 1991; Superfund Technical Assessment and Response Team, 1998; URS Corporation, 2000). The Blood Pool, so named for the deep red color of its highly concentrated iron and sulfate liquids, is a 40-foot in diameter, three feet deep, unlined natural depression that collects seep water and some surface run-off (Marcy et al., 1994; Superfund Technical Assessment and Response Team, 1998; URS Corporation, 2000).

In addition to the Midnite Mine, DMC owns 820 acres in Stevens County near Ford, Washington. The mill site complex occupies nine of the 820 acres and consists of eleven office, storage and processing (crushing and transfer) buildings. Storage pads, where uranium ore was stockpiled while waiting for processing, consume another 14 acres. The four tailings disposal areas occupy an additional 133 acres. The other 664 acres remain undeveloped and consist largely of open, mature pine forest. The property boarders Chamokane Creek which flows into the Spokane River, and is adjacent to, but not on, the Spokane Indian Reservation. Approximately 500 people live within six miles of the mill site (Washington State Department of Health, 1991).

The mill site and ore hauling route are not part of the Superfund site. Key informants reported, however, that decision-makers and community members view this distinction differently. State and federal agency representatives view the Midnite Mine and the DMC mill site as completely separate entities. In fact, the EPA has no authority over the mill site. The Washington Department of Health is the lead regulatory and managing agency for it. The Spokane Tribe of Indians and the residents of the Ford, Washington area view the Midnite Mine and the DMC mill site as highly interrelated. Not only do they share interests in both but they participate in activities concerning the management of the mine and mill sites largely as one group. While recognizing differences in lead authority, the state and federal agencies involved with both the Midnite Mine and the DMC mill site frequently interact. Hence, this project examines concerns that community members, industrial entities and environmental regulators have about both the Midnite Mine and the DMC mill site, but treats the mine and mill as one interrelated site since many of the same people are involved with the management of both sites.

DMC's Environmental Legacy

The first record of potential problems from mining wastes associated with Dawn Mining Company (DMC) operations was the observation of seepage and a white precipitate at the base of the waste piles in 1978. The matter contained aluminum salts, gypsum, approximately 3.5% U_3O_8 (uranium oxide), and high concentrations of heavy metals (Sumioka and Dion, 1983; Sumioka 1991). In efforts to contain the seepage, the DMC constructed the Pollution Control Pond immediately below the waste piles (Shepherd Miller, Inc., 1991; Sumioka, 1991). However, the seepage flowed into the mine drainages and eventually into Blue Creek. Shortly thereafter, water samples indicated that the manganese and sulfate concentrations of Blue Creek exceeded federal drinking water standards, and uranium and zinc levels exceeded uranium mine effluent limitations (Sumioka and Dion, 1983).

In 1985, four years after mining ceased, investigators noticed that Blue Creek had an unusual milky color. Upon further inspection, they detected a white precipitate on the creek sediment containing high concentrations of heavy metals (Scholz et al., 1988). More specifically, cadmium concentrations of water samples collected below the mine drainage ranged from 1.46 to 2.29 ug/l, well above the EPA chronic toxicity criteria for aquatic life (0.5 mg/l). The cadmium concentrations of fish samples were so high that investigators warned fish consumption may not be safe, especially for children (Scholz et al., 1988). While manganese, nickel and uranium concentrations below the mine drainage entrance were all significantly higher than those above the mine drainage tributary, investigators concluded they did not pose a health hazard at that time (Nichols and Scholz, 1987). Between 1985 and 1987, investigators also observed a 62% reduction in the rainbow trout population downstream from where the mine drainages entered Blue Creek. Other fish species impacted by the mine seepage include: 1) walleye (*Stizostedion vitreum*); 2) yellow perch (Perca flavescens); 3) largemouth bass (Micropterus salmoides); 4) smallmouth bass (Micropterus dolomieui); 5) black crappie (Pomoxis nigromaculatus; and 6) kokanee salmon (Oncorhynchus nerka) (Scholz et al., 1988). As a result of these concerns, two key informants reported that the Spokane Tribe of Indians closed fishing in Blue Creek for two years to protect existing stocks.

DMC Superfund Waste Issues

Spring precipitation for 1997 was the highest on record, reaching 25 inches or roughly more than five inches above average (URS Corporation, 2001b). At that time, one key informant reported that Pit 3 contained over 500 million gallons of water, nearly exceeding its back wall, and Pit 4 held more than 70 million gallons of water. With increased concerns about water breaching the pits and limited resources for remediation, the Spokane Tribe of Indians pursued Superfund assistance. Dewatering and backfilling both Pits 3 and 4 are among the environmental management challenges at the mine site. Other potential sources of environmental contamination include: 1) the abandoned protore and ore piles; 2) the mining spoils disposal area; 3) previously backfilled mining pits; 4) the Pollution Control Pond; and 5) the Blood Pool. These sources and the hazardous substances contained within them potentially pose significant threats to surface and ground water and thus, are the targets of mitigation.

Non-Superfund Waste Issues at the Mill Site

While not included in the Superfund site, management of the mill complex, especially the tailings disposal areas, is a concern for federal and state regulators, and community residents. One of the challenges is identifying the parties responsible for cleaning up the tailing disposal areas. The Dawn Mining Company (DMC) claims the former Atomic Energy Commission (AEC) contract makes the federal government responsible for any problems and needed remediation of Tailing Disposal Area (TDA) 1 and 2. The U.S. Department of Energy (which replaced AEC) takes the position that the tailings in TDA 1 and TDA 2 are not the sole product of federally contracted production, but also the result of non-government contracts, and thus, they are not responsible for financially contributing to management of the tailings (Washington Department of Social and Health Services, 1989). These two unlined impoundment areas cover a total of 59 acres and are capped with two feet of alluvial sand and gravel, and one foot of wood chips. However, the seepage of tailing contaminants into the upper Chamokane Basin aquifer that lies directly beneath the tailings disposal areas requires stabilization and ongoing monitoring. The United States General Accounting Office's position on this matter more closely sides with the Dawn Mining Company (DMC). More specifically, they stated:

> The most significant factor in favor of providing federal assistance in cleaning up commingled tailings pertains to the role of the Federal Government in creating the mill tailings situation. The mill owners apparently acted in good faith in carrying out their responsibilities in meeting various contract provisions and legal obligations . . . Now, the Federal Government is requiring the mill owners to clean up all of the tailings, including those generated before the hazard was recognized. In our view, it is unfair for industry to bear all of the costs in cleaning up the tailings . . . These are the tailings for which the Federal Government has a strong moral responsibility (Washington Department of Social and Health Services, 1989, p. 1-25).

However, there is no agreement about how to best manage of TDA 1 and TDA 2. Similarly, TDA 3 is unlined, capped with five to ten feet of fill, and topped with native grasses. Management of seepage from tailings contaminants into the Chamokane Basin aquifer beneath TDA 3 also remains unresolved. As regulators argued:

> The number of alternatives in engineering techniques is constrained by the fact that the mill has been operating for about 20 years and by the fact that accumulated tailings have already affected the environment (Washington Department of Social and Health Services, 1980, p. 6-1).

With respect to all four of the tailings disposal areas, the radioactive materials license granted to DMC by the state of Washington stipulates that:

Under the terms of the Washington State Mill Licensing and Perpetual Care Act of 1979, much if not all of the project site will be perpetually removed from alternate usage regardless of project implementation by the merit of its immediate proximity to a tailings disposal facility (Washington Department of Social and Health Services, 1980, p. 10-1).

Said in other ways:

The permanent removal of 133 acres of land from future use is unavoidable. The tailings area cannot be developed for thousands of years because of the long half-lives of the radioactive contaminants that are buried there. The other unavoidable adverse impact of the closure is that a potential source of radioactive contamination remains in perpetuity at the site (Washington Department of Social and Health Services, 1989, p. 4-44).

People must be kept away from the tailings virtually forever, or for 1,000 years. . . . To put this concern into perspective, one should not build a house or let children play on the tailings, but driving by the property or inspecting the property periodically would be acceptable (Washington Department of Social and Health Services, 1991, p. 1-59).

Tailings Disposal Area 4 (TDA 4) represents some additional challenges. The construction of TDA 4 breached the upper Chamokane Basin aquifer, with groundwater seeping 15 feet above the bottom (Washington Department of Social and Health Services, 1980). Hence, reduction in the integrity of the liner may place the tailings in direct contact with groundwater. Seepage of the tailings into surrounding soils also provides the opportunity for the natural biota to uptake potentially hazardous substances. As the resulting toxic forage vegetation enters the food chain, wildlife and human health may be at risk (Washington Department of Social and Health Services, 1980, p. 10-4). The Washington State Department of Health determined, however, these potential effects were negligible, reasoning that:

Since the life span of most animals is rather short, and populations in the wild are subjected to high attribution rates, the effects of radiation from the tailings pond would not be distinguishable from other naturally occurring forces (Washington Department of Social and Health Services, 1980, p. 10-4).

Another significant challenge associated with TDA 4 is that due to the unanticipated, premature, termination of mining and milling activities, the tailings generated following the completion of the construction of TDA 4 in September of 1981 through November of 1982 when the mill closed, occupy only 10% of its capacity. Thus, remediation activities must consider how to fill TDA 4 before capping it.

In addition to groundwater contamination concerns, the tailings areas are 800 feet from the Chamokane Creek, which enters the Spokane River six miles downstream. The annual flow of Chamokane Creek fluctuates greatly but there is a significant amount of bank cutting below the town of Ford, especially during spring flows. Seepage of tailings into Chamokane Creek creates the potential for large volumes of sediment contaminants to flow into the Spokane River (Washington State Department of Health, 1991). Woodward (1971) suggested that seepage of tailings into the creek took place as early as 1971. In a report prepared for the Spokane Tribe of Indians, Woodward (1971) suggested that the stock, agricultural and natural contamination present in Chamokane Creek would prevent untreated surface waters from meeting drinking water standards. He also reported the presence of arsenic, copper, lead, manganese, zinc and uranium in Chamokane Creek, both upstream and downstream of the mill site, and in the Spokane River. Surface water sample results presented in the mill site's draft and final environmental impact statements also indicate seepage in 1978, prior to the construction of TDA 4. However, monitoring wells were placed such that a leak 'would probably go undetected until a substantial amount of solution was released to the talus basalt aquifer' (Washington Department of Social and Health Services, 1989, p. 3-35). Stream sediment sampling from 1981 through 1987 yield:

> . . . no discernible trends. The relatively large increase in uranium during 1982 is apparently an anomaly related to sampling or analytical procedures . . . The higher levels of uranium in SW-2D water during 1985-87 may be reflected in the higher sediment concentrations of uranium at SW-2D. The effect is not seen downstream ... No impacts are seem from Ra-226 (Washington Department of Social and Health Services, 1989, p. 3-100).

With the addition of TDA 4, however, regulators concluded:

> The presence of certain heavy metals, radionuclides, and other toxic chemical byproducts of mill operation which now exceed background concentrations in Chamokane Creek should be slowed and eventually eliminated after the proposal is implemented . . . Hence aquatic resources should suffer no additional damage beyond that already realized (Washington Department of Social and Health Services, 1980, p. 12-1).

Nonetheless, given the types of wastes involved, the tailings disposal areas will most likely require some type of monitoring for many years to come.

Hazard Ranking Scores for Dawn Mining Company

Key informants uniformly reported that the following four factors contributed to delays in mitigation planning: 1) changes within agencies; 2) debates over which agency should have lead authority; 3) hopes that the uranium market would improve and interests among some to extract the remaining ore before filling the open pits; and 4) inadequate remediation funds,. Unusually wet conditions in 1997, in fact, the wettest on record (URS Corporation, 2001b), led to increased concerns about contaminated water breaching the walls of Pits 3 and 4. In response to these increased potential threats to ecological and public health, the

Spokane Tribe of Indians pursued mitigation under Superfund. The mine was proposed to the NPL on 16 February 1999 and formally listed 11 May 2000. The EPA is now the lead regulatory agency for the mine. The Washington State Department of Health remains the lead regulatory agency for the mill site. Since the mill site is not part of the Superfund site, HRS scoring only pertains to the mine site.

The HRS scoring process and evaluation of the mine site only considered surface water pathways. Ground water, soil and air pathways 'were not included because a release to these media does not significantly affect the overall site score and because the human food chain and environmental threats of the overland/flood component of the surface water mitigation pathway produces an overall site score well above the minimum required for the site to qualify for inclusion on the National Priorities List' (EPA, 2000d, p. 1). Simply put, review of previous site evaluations and an expanded site inspection conducted by Ecology and Environment, Inc. in April of 1998 resulted in a score of 100 (on a scale of 0 to100) for the surface water pathway, producing a overall site score of 50. At that point the EPA decided further HRS scoring was not necessary.

In rebuttal, the Newmont Gold Company (51% owner of DMC), submitted extensive comments about why the site should not be listed, including an alternative HRS overall site score of 24.8, slightly below the NPL threshold of 28.5. The basis of their argument was that discharges identified in site evaluations were within permitted release values. Secondly, Newmont suggested much of the data considered in the scoring process originated before 1992, prior to the operation of the on-site water treatment facility which began processing water in Pits 3 and 4, and the Pollution Control Pond, in October of 1992 (Baker and Banks, 1999, p. 14). They stated that 'this score of 50 is comparable to the HRS scores awarded to the Love Canal site [54] and the Summitville Mine site [50] . . .and higher than the NPL score accorded to Times Beach, Missouri [40],' making the score arbitrary and unfounded (Baker and Banks, 1999, p. 1). Newmont also argued that the water flows of the three drainages on the site were not adequate enough to warrant their inclusion in the surface water system. Exclusion of the drainages and the use of post-1992 data resulted in a HRS score of 24.8 according to Newmont's calculations (Baker and Banks, 1999). The score of 24.8 is below the NPL threshold of 28.5, potentially making the site an unlikely NPL candidate.

The Spokane Tribe of Indians considered the HRS score artificially low and incomplete since the groundwater, soil, and air pathways were excluded. They questioned the quality of the analytical data submitted by DMC and believed the overall site score should have been much higher. EPA defended their HRS score by stating 'the HRS does not require scoring of all four pathways, all releases or all sources at a site, if the additional scoring does not change the listing decision' (EPA, 2000d, p. 3.1-6). Since the EPA is the lead authority and they see no justifiable reason to change their position, the score remains 50 and efforts are now focused on completing a thorough site investigation in order to determine the most appropriate mitigation strategy. Site-wide contaminants identified thus far include: aluminum, barium, beryllium, cadmium, chromium, cobalt, copper, lead, magnesium, manganese, nickel, potassium, selenium, sodium, thallium, vanadium, zinc, radium-226 thorium-228, thorium-230, thorium-232, uranium-234, uranium-235, uranium-238 (EPA, 2000d, p.3.1-3). More specifically, Tables D.5 through D.15 in Appendix D describe substances identified on the Midnite Mine Superfund Site.

Interpretations of Risks Associated with DMC's Midnite Mine

Evaluation of the environmental conditions present at the Midnite Mine is not yet complete and detailed characterizations of the site are ongoing. The chief health concerns driving further analyses are the potentially adverse impacts from the metals and radionuclides thus far detected. Unlike organic compounds, metals persistent in the environment infinitely and do not, by definition, biodegrade (Allen, 1999). They can, however, transform as they migrate between environmental media and may enter the food chain when absorbed by plants in contact with soil and/or water containing metals. These features create an opportunity for metals to accumulate in multiple sources, frequently making comprehensive exposure assessments and interpretations challenging (Allen, 1999). Furthermore, most metals affect multiple organs including the digestive, cardiovascular and central nervous systems, producing both acute and long-term impacts ranging from abdominal colic to cardiac arrhythmia, renal failure, increased lung cancer mortality and irreversible cognitive deficits (Klaassen et al., 1986). The combination of substances present at the Midnite Mine makes the potential impacts to human health even more puzzling to decipher (Allen, 1999).

With respect to radionuclides, ingested or inhaled particles irradiate internal organs continuously until eliminated and given the half-life for many radionuclides, that may be extensive (Table C.8, Appendix C). The primary routes of exposure to radionuclides present at the Midnite Mine are inhalation of airborne effluents and ingestion of contaminated foods (EPA, 2000d). In efforts to estimate food chain impacts, one key informant reported that:

> We extrapolate from soil data rather than use direct tissue samples, that hasn't been done. We can get the answer we need to get from the soil samples. If anything, it tends to overestimate potential risks on the safe side. You can't clean up the deer but you can clean up the soil so its more important to focus on the source. How would you approach tissue sampling? It gets very complicated trying to determine the number of samples, different times of the year and so on. The planning would be very daunting not to mention the interpretation so we extrapolate from soil data.

Two other key informants proposed that 'extrapolation from underground uranium miners is about our best guess' of what health effects one might observe in area residents and former workers. As one elaborated, 'that can create extrapolation horrors from imagined rather than real harms but I've come to expect that.' The other informant pointed out that 'if we understood the health effects more, we would understand who is affected and how, better.' In comparison, another key informant stated that 'the human health risk is nothing like Hanford,' an active nuclear energy and weapons reservation that produced plutonium for the World War II Manhattan Project and also resides in the sate of Washington.

Newmont took a different position and claimed there is no risk to human health because no one is living on the Midnite Mine site, water on the site is not a source of drinking water for anyone, and their data provides no evidence of harm to fish or organisms in Blue Creek (EPA, 2000d). In fact, they argued, 'the data show that the environmental conditions have only improved over the years' (EPA, 2000d, p. 3.1-19). They even went as far as to suggest 'the drainage is beneficial to the fish in Blue Creek because it provides flow at times when the creek might be too low to support the fishery' (EPA, 2000d, p. 3.1-18). Hence, in their opinion, as long as no one comes in contact with the site, 'the site poses no significant risks to human health or the environment' (EPA, 2000d, p. 3.1-18). Similarly, one key informant described potential effects as 'extremely remote because there's just not a pathway for significant exposure and concentration of contaminants is not significantly above background.' For others, the issue is not that simple. As key informants explained,

> Local people have issues with the contamination, they are very connected to their ecosystem unlike most of the white population. They are concerned about the fish, hunting, plants, not just themselves. They are concerned about limiting the use of the reservation.

> The mine site is a site for hunting and food gathering. We used to get carrots and onions up there. Elk are hunted on the site. It still gets hunted quite a bit. We used to see a lot of elk, turkeys, grouse, crows and ravens up there. We seldom see grouse or ringed neck pheasants now.

> Animals love the precipitating salts around in the mine area. The area around it is open range and there has been trouble with cattle aborting. It's fenced off now but animals still get in. It would be nice not to have to worry about hunting, walking through the area, or drinking out of the stream.

> It's safe to eat only about one fish per week. That's not close to what anyone can survive on or how people traditionally live.

Another key informant reported seeing animal footprints regularly near the pits. Two other key informants reported seeing waterfowl swimming in the pits. Since there are naturally occurring uranium and radon sources in the surrounding area, however, remediation does not guarantee a solution. As one key informant put it:

> People avoid the mine area, there's recommendations to keep away from the roads in the mine area. Blue Creek has symbolic meaning and people are uncomfortable using it now. Mine related chemicals are above background but cleaning it up may not reduce them to safe levels, just background levels, just like everywhere else [on the reservation].

Concerns for local residents do not end with restrictions on land use, 'they see people dying because of the mine and mill site.' Eight key informants reported high rates of cancer (particularly lung and cervical), multiple sclerosis, rheumatoid arthritis and lupus among young adults. Trying to make sense of it, given the small population in the area and the little that is known about potential health impacts from the mine wastes, heightens the anxiety community members' have about their personal health. As one key informant reported,

> Everyone was concerned about what it would do to us, we didn't know. I've known so many people with [multiple sclerosis] and cancer. We didn't know what was causing it but we are too small to really study it.

Another challenge associated with determining human health risks stems from the fact that 'people have used tin and plywood off the old buildings at Sherwood. The company invited people to take it when they were closing things down.' Another key informant reported that 'rocks from the mine have been used for decoration and building materials like chimneys and foundations.' It is possible that 'the risk posed may be small relative to the fear.' At the same time, the extent to which such materials are used isn't clear, raising questions like 'Do we have a screaming emergency or a few places involved? How do we approach this?' For the time being, it is not clear how to incorporate this concern into the risk assessment.

Interpretations of Risks Associated with DMC's Mill Site

The tailings impoundment areas are potentially problematic in a number of ways. First, the radiation concentrations at the mill site are 2 to 3 time that of the earth's crust. One way to interpret such radiation risks is in terms of absorbed dose, or the amount of energy deposited in a tissue mass or organ. The average, estimated, background, total-body absorbed radiation dose for the state of Washington is 70 millirems per year per person. The average, estimated, absorbed radiation exposure at the mill area, however, is 173 millirems per person, per year (Washington Department of Social and Health Services, 1980, p. 2-19). Another way to think about radiation risk is in terms of activity, or time-specific decay rates relative to radium. In this context, the U.S average daily intake of uranium is approximately 1 pCi/day. Residents nearest to the mill, however, could have been ingesting as much as 50pCi/day, potentially resulting in an excess of 15 bone sarcomas in one million people. The average daily intake of radium 226 is also approximately 1 pCi/day, which is similar to area residents (Washington Department of Social and Health Services, 1991, 1994).

In addition to total-body radiation exposure, the Environmental Protection Agency's 'Radiation Protection Standards for Normal Operations of the Uranium Fuel Cycle' considers radiation exposures to specific organs. These standards stipulate that 'total doses to any organ of an offsite individual are limited to 25 millirems/year, excluding contributions from radon-222 and its radioactive daughters' (Washington Department of Social and Health Services, 1980, p. 4-8). While projected doses for the resident living closet to the tailings area were less than 25 millirems/year, the bone dose for the garden and fence post workshop area were 26.5-30.7 millirems/year. Projected lung dose for the fence post and workshop areas were 57.4-62.6 millirems/year and 36.4-37.3 millirems/year, respectively (Washington Department of Social and Health Services, 1980, p. 4-13). Since there are no known residents in these areas, this was not a foreseen concern. With respect to worker exposures, regulators concluded that:

> Except for a few individuals, the combined exposure of an average worker to [ore dust, uranium oxide and radon] over a one-year period probably does not exceed 25% of the total permissible exposure (Washington Department of Social and Health Services, 1980, p. 4-14).

Although permissible radiation exposure limits pertain only to humans, it is assumed such limits are conservative with respect to other species. Thus, as long as the mill remains within safe radiological protection limits for humans, 'no adverse radiological impact is expected' for other species (Washington Department of Social and Health Services, 1980, p. 4-14). However, plants readily absorb radium 226 and lead 210 present in soil. Wildlife may feed on this vegetation, potentially increasing an animal's ingested dose well above that of humans, to toxic levels. As this toxicity works up the food chain the contaminant concentrations may increase. Hence, since humans interact with the surrounding environment in ways different from wildlife in the area, one cannot assume wildlife and humans will experience similar outcomes. Likewise, cultural differences in how tribal and nontribal members interact with the environment may produce a wide range of ingested and inhaled doses of contaminants (Harper et al., 2002). These differences must be considered, rather than assumed absent.

When evaluating impacts to water and aquatic life, investigators concluded that the low levels of total uranium (0.002-0.011 pCi/g), thorium 230 (-0.003-1.0035 pCi/g), thorium 232 (0-0.0019 pCi/g), and radium 226 (0.001-0.01 pCi/g) detected in fish tissue samples upstream and downstream of the mill site were 'below limits for unsafe concentrations' (Washington State Department of Health, 1991, p. 2-17). Similarly, investigators determined surface water contamination impacts 'are small and far below any level of health concern and well within standards' (Washington Department of Social and Health Services, 1989, p. 3-90). Groundwater below the tailings area 'exceeded the drinking water standards for eight parameters [uranium, sulfates, total dissolved solids, magnesium, nitric acid, manganese, calcium, lead];' however, investigators reported that this water 'is not accessible for human use' (Washington Department of Social and Health Services, 1989, p. 3-95) and is not of concern since it is not being used as a drinking water source (Washington Department of Social and Health Services, 1980, 4-2). The uranium concentration was 326 pCi/l, exceeding the EPA standard of 300 pCi/l (Washington State Department of Health, 1991, p. 1-42) but according to investigators 'there are no levels of radioactivity in the accessible environment around the DMC site which would cause a health concern' (Washington State Department of Health, 1991, p. 3-97). Overall, investigators concluded:

> No levels of radioactivity in the accessible environment around the DMC site would cause a health concern . . . With the notable exceptions of uranium at the seeps and SW-3 (downstream), radioactivity in the environs of DMC attributable to mill operations are generally declining (Washington Department of Social and Health Services, 1989, p. 3-101).

In addition, the use of acidic chemicals leached uranium from the ore. Those that remain include sulfates, chlorides, and phosphates. Other heavy metals, silicates, and asbestos and PCB's from the mill buildings are also present (Washington Department of Social and Health Services, 1989, p. 1-13). More specifically,

> Windblown particles from the tailings impoundments and the ore pads also contain a variety of radionuclides and toxic chemicals such as lead, arsenic, and molybdenum. . . . In summary, uranium processing wastes contain both radiological and chemical contaminants as solids, liquids, gases, and particulates. Such processing eventually results in the contamination of building surfaces, surface soil, subsoil, groundwater and pipelines, as well as producing radon in the mill structures (Washington Department of Social and Health Services, 1989, p. 1-18).

Elevated chromium levels in surface water are another potential concern. The standard for chromium in water is a four-day average of 11 ug/l no more than once every three years, and 16 ug/l for one hour no more than once every three years. Downstream of where seeps enter the Chamokane Creek, however, four-day samples ranged from 120-370 ug/l and one-hour samples ranged from 980-3100 ug/l (Washington State Department of Health, 1991, p. 1-42). Chromium levels in fish were 'high compared to normal chromium levels in fish tissue, but following uptake by fish 99 percent of the hexavalent chromium [the toxic form of chromium] is converted to trivalent, which is not carcinogen' (Washington State Department of Health, 1991, p. 1-56). That being the case, the impact of chromium to human health was considered negligible (Washington State Department of Health, 1991). While there were no detectable concentrations of pesticides or PCB's, volatile organic, trace levels of semi-volatile organics, other contaminants of concern are described in Table D.16, Appendix D.

In addition to the health concerns area residents raised about the Midnite Mine discussed earlier, they 'have expressed concern for what they feel is an unusually high rate of cancer for such a small population,' directly associated with the mill site (Washington Department of Social and Health Services, 1989, p. 3-35). One area resident had complied a list of 25 people with cancer and suggested lung cancer was the most common form among them (Carollo, 1987). Another area resident living adjacent to the tailings area raised concerns about increased personal risk of cancer, claiming that the 'surface water from the site had washed out his road repeatedly . . . causing Dawn to install a culvert so that water would bypass his home' (Washington Department of Social and Health Services, 1989, p. 3-27). Regulators did evaluate age-adjusted death rates for the period 1983 to 1986 for Stevens County and found:

> All cancers show an increase in the county. Compared to the state total, Stevens County has higher rates of cancers of the digestive organs and peritoneum, and malignant neoplasms of the respiratory system . . . lower rates of leukemia and neoplasms of lymphatic and hematopoietic tissue . . .higher death rate for acute myocardial infarction, other diseases of the respiratory system, all accidents, but particularly motor vehicle traffic accidents (Washington Department of Social and Health Services, 1989, pp. 3:103-104).

At the same time, they concluded that 'the feasibility of an epidemiologic investigation of the health effects of low-level ionizing radiation is questionable' due to 'high costs, long disease onsets requiring extensive study time frames, and large sample sizes needed to evaluate confounders' (Washington Department of Social and Health Services, 1989, p. 3-104). Other arguments made against conducting such epidemiologic health studies were that 'it would provide no major information pertinent to the closure effort . . . expected risks are low' and 'if radon emanation can be kept below the standard, the population is adequately protected' (Washington State Department of Health, 1991, p. 3-99). For this reason, the Washington Department of Health and other agencies have not completed additional human health investigations. This position has not alleviated community concerns. As one key informant put it:

> The state would say 'it won't hurt you' but how do you know? They don't know what it will do. It's [the land for the mill site and the mine] never going to be able to be used for anything. You don't know what's going to happen. From my house you could hear the pumps. That's the thing that bothered us the most is they don't know what's going to happen down the line.

Mitigation Goals for DMC's Midnite Mine

Mitigation planning will not commence until the completion of site characterization. Key informants did, however, express ideas about mitigation goals and challenges. One of the challenges will be the maintenance of the groundwater treatment facility on the mine site. It is not clear how long groundwater treatment will need to continue and since it costs over one million dollars a year to operate as one key informant projected, no one is interested in assuming its responsibilities. Another challenge is 'reliance on man-made barriers where there are large amounts of toxic materials that have to be contained indefinitely.' Since the hauling route was recently re-paved, identification and remediation of spills may also be problematic. With respect to Blue Creek, one key informant recognized 'continuing contamination of the creek can only be controlled, not eliminated,' making its long-term management important.

Determining how to backfill the open pits and the composite of adequate surface cover will also require careful planning. As one key informant suggested,

> The state is not taking a conservative position to put in an adequate cover. They're setting up conditions that will duplicate conditions at Hanford where it's taking millions of dollars every year just to keep it partially under control. An adequate barrier would be no less than 17 feet like at the Sherwood site, [about two feet more than proposed].

Five key informants posed the question, 'why is it taking so long to cover up the site? It didn't for Sherwood.' Another asked, 'why isn't the company being held responsible? We have nothing to show for it.' Two additional key informants expressed an interest in wanting 'to walk across the landscape and not be harmed by it, to have sweat lodges at the base of the mine again, and to eat fish, deer, elk, and berries without having to worry about it.' Another key informant 'would like to see it exactly the way it was before the mine was there.' But as another explained, 'it's like taking a glass egg and cracking it open. How do you go back to a glass egg once it's split apart? Lost opportunity costs come with mining.' Nonetheless, 'the tribe would like to see the site cleaned up and restored. EPA is tending towards much less clean up than we would like to see. They'd like to put up a fence and call it good.' The tribe does 'get some funding for clean up and a fish study, that supports a few people' but that does not adequately compensate for the permanent loss use of land. Clearly, there are many difficult decisions to make ahead.

Mitigation Goals for DMC's Mill Site

One of the most significant mitigation challenges at the mill site involves tailings disposal area 4 (TDA 4). Mill processing wastes occupy only 10% of TDA 4's capacity. Hence, closure of TDA 4 will require adding large volumes of fill material prior to capping and final surface area restoration. An avenue pursued to obtain the massive volumes of fill materials needed was the federal Formerly Used Sites Remedial Action Program. This Nuclear Regulatory Commission program seeks ways to dispose of low level contaminated soil, including radioactive mill tailings (Washington Department of Social and Health Services, 1989). Under this program, imported waste materials can not be more radioactive than conditions already present at proposed disposal locations. Fill materials proposed for TDA 4 under this program were 'select material' containing 'low level, naturally occurring radioactive material needing a disposal site, but having a level of radioactivity too low to be considered acceptable at already established low level waste sites' (Washington Department of Social and Health Services, 1980, p. 6-3). The basis of the argument for using select material in TDA 4 was twofold. First, 'since the present disposal site is already contaminated it makes better sense to confine future disposal to the same site rather than to contaminate any other area without strong cause' (Washington Department of Social and Health Services, 1980, p. 6-3).

Secondly, any revenues generated from importing select materials would assist remediation efforts at the Midnite Mine.

The community reaction to the proposal of importing select material was very negative. Community members opposed to the select material formed a group called Dawn Watch. One key informant reported Dawn Watch had over 100 members, both local and nonlocal people, at the height of the controversy. The biggest challenges Dawn Watch faced initially according to one key informant was drawing attention to the site, in part because 'the mill property and even the mill ponds are barely visible from the road. Most people were not even aware of the location of the site.' Other key informants suggested that 'the site is out of the way so it doesn't get as much attention as it should' and 'we're a small population, if we were bigger like Mercer Island we would have gotten a lot more attention.' At first, one key informant explained:

> The Sierra Club and Green Peace wouldn't help us, they turned us down flat. But I didn't want to see anything happen to my children. We visited with everyone we could think of to draw attention to it. The same few people did all the work and remained through the whole process. We put a lot of time into it, long distance telephone calls to environmentalists, stamps and letters to senators and the president, meetings, I took time off of work, we were always thinking about it.

Another key informant reported that:

> The strangest thing about the Sierra Club is that I personally contacted them about the Dawn issue at the very start and was told 'that issue is too small for us to deal with.' Only when we had stopped Dawn in their tracks did the Sierra Club show any interest and of course by then the news had it nationwide. Good publicity hounds!

Grounds for the opposition stemmed from three primary factors. The first factor was little confidence in the company selected to manage the TDA 4 fill imports. One key informant reported that:

> Envirocare out of Utah was selected to head up importing the dirt. They had over 60 infractions filed against them. How dare you tell us what we as a community can and can't do, when they can't do it right! How dare Utah tell Ford, Washington what to do!

Another primary factor was whether Ford, Washington, should or should not be responsible for housing wastes generated elsewhere. As one key informant explained,

> We didn't want to bring in anymore than what was already here. Why should we be responsible for New Jersey or New York or some one else's waste? They should find their own local place for it like we did. Why should we be subjected to what some one else did?

Elaborating further, another key informant stated:

There was also the question of environmental genocide. I know that is a heavy name for it. But the way I see it countless times the powers of this nation have been all too happy to shuttle toxic and nuclear waste off to areas where Indians or poor people live. Many of these locations have been unorganized politically and thus unable to fend off the incoming waste . . . often they are saddled with the consequences of previous, uninformed decisions by people who are now dead or gone. Midnite Mine is a good example of that. It is not a very long stretch to see eastern, urban, NIMBYs consciously or unconsciously deciding it was okay to send these harmful materials away to a place they didn't care much about--a poor community out in the woods, or better yet a poor Indian reservation near a poor community out in the woods. Its just too bad if it adversely affects the people's health or even kills them.

The transportation route of the select materials was the third primary concern. The road selected was not adequate to handle the traffic volume or weight of the trucks projected to 'be going through Reardan every 3 minutes,' requiring at least $5 million worth of road improvements. The site of the proposed terminal was within a quarter mile and upwind of a school yard, raising additional concerns about toxic dust exposures and anticipated accidents that the volunteer fire department may not be able to respond to. Other options involved offloading from trains in Springdale, Loon Lake or Deer Park and hauling it by truck from there, but those roads were also inadequate.

As part of the process to secure a bid for select material and alleviate community concerns, the Dawn Mining Company (DMC) formed the Local Citizens Monitoring Committee (LCMC). The purpose of the committee was to establish criteria for fill material, including transportation, handling and storage procedures. Only select materials meeting all of their requirements would be disposed in TDA 4. Among the twelve applications received, the nine people that resided within five miles of the mill site were selected for the LCMC. The minimum criteria stipulated for the select material include: 1) only radionuclides with decay chains of natural uranium and thorium-232, i.e., radium-226; 2) the maximum concentration of radionuclides in imported materials can not exceed that of the tailings already present; 3) an average radium-226 concentration limited to 100 pCi/g or less; and 4) no non-radioactive materials that would result in a classification of 'hazardous waste' by RCRA definition (Washington Department of Social and Health Services, 1989, p. IV-1). In contrast, clean soil standards are no more than 5 pCi/g for radium 226 in the top 15cm and 15 pCi/g for soils deeper than 15cm (Washington Department of Social and Health Services, 1991, p. 3-90).

One source of 'select material' considered was ash from the Spokane incinerator but its unknown volume, content and availability eliminated it as a viable option. Another source proposed by the Dawn Mining Company in December, 1985, was New Jersey dirt. This dirt, stored in 55-gallon drums, consisted of soils excavated around the foundations of houses built on former sites of radium processing facilities, associated primarily with radium watch dial painting, and some pharmaceutical wastes. The excavated dirt is stored in 55-gallon drums as it awaits disposal (Washington Department of Social and Health Services, 1989, p. 1-57). The Washington Department of Health concluded that the New Jersey fill dirt 'presents no statistically significant increase in human exposure to radiation over normal mill operations or over background radiation' (Washington State Department of Health, 1991, p. ii). They also recommended the import of the New Jersey dirt on the grounds that it would help offset mill site and mine closure costs but asserted 'the department will not make a decision on this alternative until public comments are received on this draft supplement, and until completion of a final supplemental [Environmental Impact Statement]' (Washington Department of Social and Health Services, 1991, p. ii). Another location briefly considered for the New Jersey dirt was the Teledyne Wah Chang Superfund Site in Oregon (discussed in chapter five) but Oregon radiological waste laws prevented its importation (Washington Department of Social and Health Services, 1989, p. V-21).

The average radium-226 concentration of the New Jersey dirt, however, was 'about 300 pCi/g' with 86% having radium-226 levels below 200 and averaging 40 pCi/g, 12% ranging from 200-2000 pCi/g, and 2% over 2000 pCi/g (Washington Department of Social and Health Services, 1989, p. IV-4). Concentrations of Ra226 in existing tailings, slimes and sands ranged from 157 pCi/g to 673.6 pCi/g, with an average of 260 pCi/g (Washington Department of Social and Health Services, 1989, pp. II:5-6). Since the average Ra226 concentration of New Jersey dirt exceeds that of the present tailings and EPA criteria, its use was the subject of great debate. As one key informant put it,

> There was the issue of the misrepresentation of the materials to be imported. The company stated that the material was so safe that you could take a bath in it and yet failed to mention that test holes at the source sites showed that it contained various levels of non-nuclear toxic waste, including PCBs, DDT, DDE and many other known carcinogens and mutagens.

From another perspective,

> [The Local Citizens Monitoring Committee (LCMC)] determined the level, set it to be below background, no worse than what was already out there. We had the right to check the material, trucks, handling after it came on site, to take samples, to check for radiation, and that the trucks were clean before they left the site. . . The real thoughts of the LCMC were if we made it tough enough, we thought no company would give them the contract. We had the right to monitor and take split samples in New Jersey. We felt secure there weren't many companies that would honor our rights. During all of this, DMC management changed three times. First management agreed to bring [low level waste] in, in sealed bags. New management said loose dirt in covered trucks was O.K. and four members walked out of the meeting. With DMC consent, we looked at different sealed containers. The ones demonstrated split open when they were put into the pond. We said no and finally found one that we approved. But DMC never got a contracted and changed to clean fill.

The draft supplement to the Final Environmental Impact Statement (Washington State Department of Health, 1994) discusses DMC's submission of an alternative plan to the selected clean fill dirt option in which they once again, proposed the New Jersey dirt as alternative fill. Twenty-two months later the Washington Department of Health decided to revisit this alternative on the grounds that DMC did not have funds to remediate the mill site and excess revenue could be used to clean up the Midnite Mine. In fact, the state approved a modification of DMC's radiological waste license to house the New Jersey dirt in TDA 4, largely in the opposition of community desires. The message to the community was if they wanted anything cleaned up, they would have to accept the import of out of state wastes. In response, Dawn Watch refers to this ultimatum as blackmail.

The debate over the '"New Jersey" dirt has mobilized the community to look suspiciously at any closure effort by the Dawn Mining Company' (Washington Department of Social and Health Services, 1989, 1-58). Even after offering the community incentives like building a baseball field, restoring the old Ford schoolhouse, expanding the fire services and educational facilities after the mill closed, and twelve years of negotiations, DMC was not able to secure a bid to import 'select material.' Instead, a key informant reported that 'an unlicensed facility in Idaho, they aren't required to be licensed by the state in Idaho, was selected. They underbid [the Dawn Mining Company] and are making quite a large profit.' When all was said and done, the primary closure activities of the mill site include: 1) solution removal of liquids in TDA 4 (i.e., 100 million gallons); 2) groundwater remediation; 3) removal of mill complex buildings and surrounding contaminated soil, with all demolished and extracted materials disposed in TDA 4; 4) fill TDA 4 with 4.1 million cubit feet of clean fill material; 5) stabilization of TDA 1 and 2 (20 million cubit feet), and TDA 3 (33 million cubit feet); and 6) final covering of all ponds and other windblown contaminated areas (Washington State Department of Health, 1991). Even under the best circumstances, however,

> This long-term plan can only be considered tentative because: (1) no one has experience with long-term monitoring of uranium tailings sites; and (2) the plan assumes that no significant geological or hydrological changes will occur and that the cover will maintain integrity. If problems with the cover develop or data show increases in contaminants from the tailings, the monitoring would be increased and mitigation actions taken as necessary (Washington Department of Social and Health Services, 1991, p. 2-38).

Preliminary Conclusions

As we learned in chapter four, the community surrounding the Dawn Mining Company (DMC) Midnite Mine and Ford mill site defines itself largely in terms of how its members interact with bio-physical landscapes. Within this symbolic context, the area suffers from high, chronic unemployment, and it is common for area residents to commute to Spokane to work and shop. Such a dependence on external ties for daily survival takes time away from local associations, making it difficult for community members to influence local decision-making activities (Freie, 1998). Due to time constraints and a lack of resources, community members may find themselves relying on technical experts and environmental regulators in new ways (Chess et al., 1995; McGee, 1999).

The need for external funds to mitigate the Midnite Mine introduces additional outside agency personnel into an already disjointed decision-making process. Given the general suspicion community members have of outsiders and a long-standing distrust in government rooted in nonconsensual land use agreements, the addition of more outsides controlling local decisions complicates the ability of community members to engage in decision-making processes even further. This is not to suggest community members are not interested in influencing local decisions, but rather the ability to do so comes with constraints. Long-term and close connections to the land make having a say in local land use decisions important for Wellpinit and Ford community members. Key informants reported that most people have lived in the area for a long time, especially those most actively involved in decision-making about the mine and mill. As one key informant put it, 'I've been here 28 years but I'm a newcomer to some.' Just the same, multiple networks, privacy desires and the lack of extensive local resources promote individual rather than group responses to risks. However, community members 'rally together in times of need.'

The Dawn Watch group is a good example of how community members pull together in response to a commonly shared concern. Dawn Watch was a coming together of tribal and nontribal community members for a shared and very specific cause—to prevent the import of the New Jersey dirt. Once accomplished, many people returned to focussing on their individual needs. Community members also successfully acquired Superfund resources for remediation of the Midnite Mine. While there is little evidence of long-standing community-wide organizations outside of the tribe, Wellpinit and Ford, Washington community members are not without power but rather selectively exercise and organize their different groups for specific interests. The ability to formulate common courses of action despite the differences that may exists between groups may be the product of some community members giving in on less important issues in exchange for having their way on more important issues (Coleman, 1986). The mutual dependence on their landscape also provides a common ground to develop responses to specific interests. For Wellpinit and Ford community members, specific interests build cohesive relationships rather than reduce cohesion among community members.

External organizations may underestimate the ability of the Wellpinit and Ford, Washington, residents to organize responses, given their rural and individualistic character as well as their dependency on external resources, especially for remediation of the Midnite Mine. At the same time, delays in reclamation, not seen with the Sherwood Mine and Mill, frustrate community members and make them feel like they have little influence over external agencies' actions. The declaration of decision-making authority by external agencies reinforces this sense of powerlessness. Since experts leave many of the community members' questions unanswered, there is a sense that expert knowledge offers little to the solution and that experts may be withholding information. Little confidence in experts may make it difficult for experts to dominate the policy process. Federal mandates specifying the role of external agencies, however, may override equitable opportunities to participate in decision-making. Lack of media attention may also help keep everyone's interests on the same playing field. At the same time, lack of media attention may reinforce community members' frustration with being unable to draw attention to local issues that are important to them.

Adding to the uncertainty of community members' influence in local decisions is the fact that the health effects associated with the DMC's Midnite Mine and mill site are not clear. In part, little is known about the specific health and ecological outcomes associated with the mixture of contaminants that are present at the mine and mill site. Secondly, characterization of contaminants present at the mine has begun only recently despite their persistence for multiple decades. Thirdly, who is responsible for the wastes contained in tailings disposal areas 1 and 2 at the mill site is under debate. While this uncertainty and lack of confidence in information provided by experts may increase fear of potential hazards for some, others argue cause for concern is not necessary as long as neither humans nor animals interact with the substances present. Moreover, a significant underlying challenge to controlling the wastes is that the area has many natural sources of radon emissions, making it impossible to eliminate risk completely. At the same time, those that expressed concerns about increased cancer, rheumatoid arthritis, lupus, and multiple sclerosis among the area's young adults in particular, suggested their concerns are not being taken seriously while also acknowledging such issues are difficult to study in a population as small as theirs.

While disruption from historical land use and cultural loss remains evident, there is no clear mass exodus as a result of concerns related to the mine and mill. In fact, one key informant reported 'people are moving in all the time, there's a lot of good benefits for reservation residents.' On the other hand, the only place tribal members can practice traditional lifeways is on the reservation. This requires tribal members to remain connected to even contaminated land that cannot be used. From another perspective one community member stated, 'I like having that thing up here. If it keeps Spokane people away from this beautiful land, then that's good' (Walter, 1989). Just the same, one should not infer community members have become at ease with the risks present.

Perhaps the devaluing of local knowledge draws attention to the differences among community members. The subsistent-based lifestyle of many community residents involves a different relationship with the land, which may inadvertently increase their exposures to substances as such hazards enter the food chain (Harper et al., 2002). The failure to understand this lifestyle will inhibit the identification of all potential exposures, especially if outside experts underestimate wildlife access to the area and how imposed land restrictions as a measure to minimize bioaccumulation inhibit cultural practices. Failure to acknowledge these differences promotes further debates, reduces cohesion between environmental regulators and community members, and encourages dissention. It also highlights the fact that the structure of the decision-making process for this Superfund site appears to primarily inform, rather than actively involve, community members. In fact, community members appear to have very little control in defining the structure of decision-making, given the clearly defined lead authoritative roles of outside agencies. Since mitigation activities are likely to extend across decades, overlooking these important issues now may make the road ahead difficult.

In this small, isolated community where economic resources are minimal and agreement about issues is not uniform, the community is dependent on outside resources to minimize the risks under debate. The failure of experts to clearly answer their questions, questions that arguably do not have definite answers, exacerbated the long-standing distrust in regulatory agencies and DMC. While there is no mass exodus from the area, the community's chronic unemployment and lack of resources make members vulnerable to the decisions of outside agencies, like external agencies attempting to impose the import of hazardous substances generated elsewhere into their community in order to fund remediation efforts at the Midnite Mine.

Uncertainty is not, however, limited to potential health effects. It extends to who will be responsible for long-term management of the risks. For example, the water treatment facility at the mine site cost roughly $1 million dollar per year to run. This same community does not have water and sewer infrastructures in part due to a lack of funds to operate them. Hence, local resources to operate the mine's water treatment facility do not exist. Furthermore, experts specified that the mine and mill site areas must be monitored for at least 200 hundred years. This raises additional concerns about economic resources needed for long-tern risk management as well as the ability of everyone to remember the location of the hazards, given the indigenous trees and grasses planted on top of them, not to mention the little confidence local people have in environmental regulators.

What we can learn from this case this case, however, is that creating opportunities where communities maintain some sense of local control in mitigation decisions, largely made by external agencies, is essential for minimizing conflict and encouraging community involvement in decision-making about environmental risks. If community members do not feel they have any influence in decision-making, they have no incentive to participate. The intimate connections that Wellpinit and Ford community members have with the landscape put them in a position to offer invaluable information about important impacts and insights on how to control such impacts. This can be a tremendous asset for environmental regulators. It is also important for external agencies to be cognizant of their potential impact on local power structures and not to underestimate the power of those structures, e.g., Dawn Watch, even in areas that may be seemingly forgotten about because their rural, out of the way, locations.

Another important consideration is developing strategies to communicate uncertainty in ways community members can understand it, allowing them to make better informed personal decisions. This is especially important given the enduring nature of the hazards present in the Wellpinit and Ford, Washington landscape as the uncertainty associated with such substances will most likely remain unresolved for many years to come. That is not to suggest that decision-makers and community members should simply look beyond their differences and move on, but rather they need to recognize their differences and learn as much as possible from them in order to move ahead. One must develop realistic expectations about erasing the ingrained distrust among those involved in the decision-making process and reversing top-down risk management approaches overnight. But as one key informant asked, 'until you've read what we've done, how can you tell us what to do?' Perhaps one place to start would be less telling and more listening.

Chapter 7

Public Participation in Environmental Decision-making: What is it and When Does it Work?

The quicker one recognizes that the public will be involved the better. You just have to recognize it's going to happen and to begin dealing with it instead of kicking and screaming against it (Key Informant).

Introduction

Social interaction provides an opportunity for people to better understand situations they share, including their physical environs (Gould, 1978; Kirkpatrich, 1986). As our interactions span multiple locations and grow increasingly diverse in today's global society, however, maintaining enduring place-based bonds and commitments becomes more difficult. These interactions also transform how we think about community (Allen and Dillman, 1994; Tönnies, 1940). In fact, for some communities social interaction may play a more significant role in defining community structures than physical features (Doreian and Stokman, 1997; Rifkin, 1995; Schuler, 1996). Under such circumstances, participation in shared situations is particularly critical for formulating and maintaining community structures as well as cohesive relationships among community members (Etizioni, 1996; Nisbet, 1990; Selznick, 1992).

In the midst of social transformations, some community affairs may be exceedingly difficult for community members to participate in. For example, when outside entities make and enforce local decisions, a lack of time and resources to participate, coupled with a sense of little control in decision-making, may discourage community involvement (Chess et al., 1995; McGee, 1999). At the same time, decisions may be difficult to reach when community involvement is sparse (Vorkin and Riese, 2001). This makes social interaction among community members important for both defining and redefining community structures, and community-level decision-making. In order to better understand how participation shapes community structures and community-level decision-making, this chapter examines how community members, industrial entities and environmental regulators participate in decision-making processes about the Teledyne Wah Chang Albany Superfund Site and the Dawn Mining Company's Midnite Mine Superfund Site and mill site. Specific items of interest include lines of authority, frequency of participation, factors that decrease participation, factors that increase participation, benefits from participating, and ways to improve participation.

Lines of Authority in Millersburg and Albany

Lines of authority with respect to Superfund decision-making are clearly drawn among environmental regulators, industrial entities and Millersburg and Albany residents. Uniformly, key informants described decision-making about TWCA as 'an issue for [the Environmental Protection Agency (EPA)], [the Oregon Department of Environmental Quality (ODEQ)], and the company; the community isn't really involved.' Technically, 'the EPA is the lead agency, working jointly with the state,' but as much as '99% of [the monitoring] is done by ODEQ.' For this reason, key informants assert that, 'the EPA and [ODEQ] are the decision-makers,' whereas '[TWCA] is just legally required to perform the selected remedy.' In efforts to use resources as wisely as possible, the EPA and ODEQ closely coordinate risk management efforts. As one key informant explained, 'there's always loop hoops to jump through but we go to great lengths to avoid duplication and stepping on each other's toes.' This is not to suggest enforcement of mitigation actions is always straight forward. One key informant confessed that at times, environmental regulators 'have to make [TWCA] do things because the law says so but [the environmental regulators] don't always agree' such requirements are necessary.

While TWCA 'has some input,' four key informants framed the decision-making process as 'very adversarial' at times, proposing that '[TWCA] was forced into compliance,' and that they 'just have to take it.' Conversely, one key informant suggested that 'TWCA has more resources than even the state' to participate in decision-making activities, especially with respect to legal council. Generally speaking, 'EPA runs the public involvement process. [TWCA goes] along with it most of the time but they air differences in public sometimes.' Moreover, 'the community has backed the company every step of the way for more reasonable alternatives in opposition to EPA.' There has also been 'no real environmental representation' to any significant degree by outside or internal community groups. In fact, four key informants reported that there are no local environmental organizations to speak of. Others argued that 'the state listened to concerned outsiders the most' and that 'the local perspective is not well represented in the broader context.' With respect to Superfund issues, however, all key informants reported that the EPA has the final say.

This suggests a number of things. First, community members may interpret the top down management style and mandated authority of the EPA to make local decisions, as an atmosphere in which they have little influence. That being the case, the incentive to participate in decision-making processes about TWCA is marginal. Their attachment to the family-wage jobs provided by TWCA may encourage them to have little interest in outsiders', like environmental regulators', perspectives. Job attachment may also infer that they feel they cannot leave the area and thus, they have little interest in the seriousness of environmental concerns since they have to live with it (Shriver et al., 2000). In combination with less than obvious opportunities to participate, these factors may discourage community members from participating in community-level decision-making about environmental risks associated with TWCA.

Lines of Authority in Wellpinit and Ford

Decision-making authority for Dawn Mining Company's Midnite Mine and mill site is a bit more complex than that of the TWCA site. Technically, the EPA is the lead authority for making decisions about the DMC's Midnite Mine. Prior to EPA's involvement, however, the Bureau of Indian Affairs and the Bureau of Land Management oversaw the mine site. The BIA and BLM remain involved but merely at an information exchange level. The EPA frames their relationship with the tribe as 'government-to-government' while maintaining their lead authority such that '[the EPA] won't change [its] position if it is technically correct but [the EPA] position reflects input from the community.' This suggests the EPA may consider community input for some issues more than others, and that they exercise their full authoritative control in local decisions regarding the mine in all situations. From the perspective of some key informants, the distribution of authority among participants in the decision-making process follows less equitable lines. For example, two key informants suggested:

> EPA has the hammer for the mine. It's very disproportionate, they've pushed other federal agencies aside. It's out of control and we hardly get to communicate with others involved. DMC has had almost no say with the mine. If it wasn't for the water treatment plant, DMC would be trespassers.

> There's a feeling that EPA isn't supporting the tribal position on health assessment. For example, with a subsistence type of diet people have a much higher exposure by eating the fish and plants that have contamination in them, but this isn't taken into consideration.

With respect to the DMC mill site, '[the state of Washington's Department of Health (WDOH) consults] with the tribe and local citizens but WDOH has the final say.' Others described the hierarchy of power relationships among environmental regulators, e.g., EPA, Bureau of Indian Affairs (BIA), Bureau of Land Management (BLM), and the state of Washington's Department of Health (WDOH), and community entities as follows:

> EPA is the lead. The tribe has power through being able to get senators and congressman to support the need for clean up. The role of BLM is unclear, they have trustee responsibility to clean up the site and oversee DMC about what to monitor. BLM and EPA kind of share it and the tribe is consulted.

The EPA has the most weight, the tribe next, after that the BIA, BLM and the state, they have about the same.

Until 1999, the tribe wasn't that involved and trustees made most of the decisions for them. Now EPA is the lead, BLM and BIA are not involved much now. The tribe is more involved, they pushed for Superfund status. BLM and BIA developed a reclamation plan but it went to [Superfund] to get the money needed for reclamation.

The federal agencies are squabbling with each other [about who is in charge].

It's much more even with the mill. [WDOH] encourages participation of other agencies, they're quick and open to suggestions from other agencies.

Hence, persons participating in local decision-making processes propose an important distinction between state and federal agencies; state agencies may be more open to community input than federal agencies. Given the community's historical adversarial relationships with federal agencies, this is not particularly surprising but this illustrates that community members do not frame all agencies in the same negative light.

One key informant reported that 'decisions are made by a very few number of people. They put forth their own agenda . . . but I don't know how representative they are of the general tribal population . . . [some] have more credibility than sometimes deserved.' Said another way, another key informant stated that 'some plans have been railroaded, not because the plans weren't based on good science or weren't thought out, but because it didn't advance their own agenda.' For one key informant, 'it has seemed to me throughout the process that our group was always pushing a rock uphill.' Expanding on this idea, another key informant explained,

Often, the only comments that had any weight attached to them were technical in nature and it seemed like those people with the least resources were being required to come up with the evidence. The company presented a plan that was based on the best interests of the company, not on the local population of the taxpayers who would be paying for the mill closure. The State then asked if it met the technical rules, but not the best interests of the local people, or the taxpayers. Then it became our job to wage the struggle on technical grounds; health, safety, water quality, or how the community felt about the issue was considered relatively unimportant.

Public meeting records offer further support for this claim. Twenty of the 41 comments made by community members during public hearing on 20 and 21 March 1989, regarding the Draft Environmental Impact Statement for the mill site closure were followed with 'your comment is noted' and nothing more. For example:

COMMENT: We wonder why the New Jersey legislature is willing to spend an extra $139,700,000 of New Jersey taxpayers money to ship the material out of state if it is so safe?

RESPONSE: Your comment is noted (Washington State Department of Health, 1991, p. 6-80)

Thus, from the community members' perspective, their claim that they have little influence in decision-making is supported. This may in turn discourage further participation and inadvertently tap into historical conflicts about excessive federal control in local matters.

Frequency of Participation in Millersburg and Albany

Key informants consistently reported participation in decision-making about TWCA from the community at large has been 'very little, it's pretty much between [TWCA], state and federal agencies.' The EPA held two public meetings in regards to mitigation plans as mandated but no additional meetings have taken place. Local residents 'are aware of the site's Superfund status but the state takes almost full responsibility for monitoring everything' and there is 'no public interest.' Low public interest, however, may in part be due to little opportunity to participate in decision-making processes, coupled with time constraints and competing responsibilities. As one key informant put it, 'I just don't think about this very often.' Unclear roles in decision-making processes also contribute to decreased participation. For example, one key informant stated 'as an organization we have had very little involvement or influence pertaining to the [TWCA] environmental issues . . . it is outside of our jurisdiction.' According to another key informant, the majority of the 'public interest in this site has been from [outside] environmental groups supporting the clean up and the TWC employees wanted to leave things alone . . . There are no, and have been no, work groups or task forces due to the low level of public interest in the site.' All key informants agreed that 'it's very hard to get people involved,' especially when they do not have an obvious role in decision-making and when decisions only appear to affect them indirectly.

Frequency of Participation in Wellpinit and Ford

Similarly, key informants reported that 'most of the time [Wellpinit and Ford] as a whole doesn't want [public participation] but there's always a few that try to make it better.' The key informant elaborated to explain that in part, people do not get involved because it conflicts with their desire for personal privacy and they generally distrust federal agencies. With this distrust comes the perception that community members have little influence in decision-making processes. Building on this notion, one person suggested 'if you're noisy enough they listen but it's always the same few and I don't want to do it again, I'm too old.' As this suggests, the time commitments for the few that get involved may become quite excessive, given limited resources and the few people to share the workload of such voluntary efforts. Over time, people may become burned out from multiple competing demands and in turn, reduce their participation.

At the height of the controversy regarding the New Jersey dirt discussed in chapter 6, 'there were over 100 Dawn Watch members, [a group of people opposed to the New Jersey dirt], but as the issue became more complex people became overwhelmed and pulled away.' People also found it difficult to find time to remain actively involved in Dawn Watch activities as the controversy dragged out. While 'Dawn Watch is not as interested in the mine as the mill site . . . some of [the Dawn Watch members] are tribal members.' However, 'very few people from the reservation got involved, the most interested people weren't from around here.' This suggests specific issues encourage selective participation in that particular issue, not necessarily in the community where the issue originates.

Another key informant reported that 'for the mill site meeting, there was a big turn out. Dawn Watch drew a lot of attention to the [dirt proposed for import]. For the mine, there was a moderate turn out, mainly just the tribal council.' Again, this is an issue-specific type of response. The issue behind the New Jersey dirt was about bringing some one else's problem waste into the community. This is an issue that is not unique to the Ford and Wellpinit community. Other communities face similar impositions from outsiders, especially poor and minority communities (Bullard, 1990; Been, 1994; Mohia, 1995; Pellow, 2002; Pellow and Park, 2002). By supporting this particular situation, people from other communities can support an issue that may in turn further their own interests. The mine on the other hand, has been an unresolved community-based issue, unique to Wellpinit and Ford, for decades. Outside interests have less to gain from being involved with the mine compared to being involved with the specific issue concerning the import of hazardous wastes, an issue that may be germane to where the outsiders live.

Generally speaking, a key informant stated that:

> The people here are pretty indifferent . . . people just don't care about the politics. It's hard to get people involved in anything. Nobody ever shows up to any of the meetings. They come out here to get away from everything.

'The biggest challenge' according to four key informants, 'is reaching tribal members,' but they did not have advice to offer on how to increase their participation. Another key informant observed that:

> There's been dismal attendance at the meetings. Attendees are basically the regulatory agencies and the company, with very few people from the general public. People are trying to advance their own agendas so decisions are made by a few people that may not have the community's best interest at heart.

Again, this suggests support for specific issues may be more readily fostered than support for involvement in the community per say. At the same time, 'it's hard to get a sense of what the community is like if they don't show up at the meetings.' Along those lines, one key informant reported that:

Some people attended the meetings as a form of entertainment, they came for the fun of it, the pure joy of it. For a small town, it was a big social event and some speakers enticed the audience with tales. Some came only for the cookies.

But on a day to day basis, 'the community doesn't really think about [the mine or mill] that much.' As one key informant explained, the community at large looked to the few involved 'kind of like "you take care of it, we don't want to know" and they felt like we were looking out for the welfare of the community.' This further supports the notion that people may participate in specific issues more frequently than general community matters.

Factors that Decrease Participation in Millersburg and Albany

One leading factor that decreased participation in decision-making about TWCA among Millersburg and Albany residents was the notion that 'on a day to day basis people really don't think about it.' Said another way, 'unless you drive by it and see it and have to deal with it everyday, people don't get involved.' As a nonrespondent suggested, 'it's a nonissue so far as we can tell, I haven't heard anything in our paper for years.' Hence, unless community members are directly affected by an issue, there is only a marginal incentive to become involved at best.

Within the Millersburg-Albany culture, two key informants proposed that public participation is 'viewed as an obstacle, not something to facilitate' and 'if a worker is concerned about the community's health, they are mislabeled as an environmentalist opposed to jobs.' This suggests the community norm may be to not participate in local affairs. Further supporting this normative idea, another key informant stated that 'one of the biggest problems in society today is nobody wants to get involved. You could throw rock into the auditorium at a public meeting and not hit anybody.' When people feel that 'there's not a obvious role for us' because they 'were never asked' and were 'not invited to the community working group meetings,' or the issues 'fall outside of our charter and scope of practice,' they are also reluctant to participate. Lack of resources, or as one key informant put it, 'everyone's understaffed,' makes it difficult for community members to get actively involved as well. Poor communication about when meetings take place also contributes to low attendance. As one community member reported, 'I would of showed up at the meeting . . . but I did not know about it' (TWCA Administrative Record 14.4-0000001).

Different perceptions about the role of regulatory agencies in the decision-making process may serve as a source of contention that discourages participation. For example, one key informant felt that 'regulatory agencies are as much a part of the problem as waste producers. They tend to bend over backwards to protect waste producers and aren't forthcoming . . . they seek out the least costly means.' From a regulatory perspective, another key informant expressed frustration with the fact that:

Often what we can do is not good enough for the public, they want more. Land use issues are a good example. The public may say 'I don't want that plant there' but we can only regulate companies that are already there, we can't make them move. Sometimes they get frustrated when we don't do things that we don't have the authority for . . . the public thinks we are trying to protect the company or trying to hide something. The public doesn't know everything we do to track the company. There's almost a paranoia about it. They might see something and all it is is steam. We know who the heavy polluters are, we keep leaning on them and fining them more. Companies can't get away with [being heavy polluters] in Oregon. The public and politicians won't support them.

This does not explain TWCA's ongoing operations at the site in spite of its history of environmental violations as discussed in chapter five. Inadvertently, this perceived tolerance of environmental violations may build upon established distrust in both TWCA and environmental regulators.

Disagreements among experts and environmental regulators may also reduce community members' confidence in lead agency decisions. For example, two key informants proposed that:

Most barriers involve the lack of agreement between [ODEQ] and the company on scientific information. For example, should the discharge be to a small creek that dries up each summer, or in the large Willamette River. Much work has been done to answer this question, but arguments persist about the dissemination of the gathered data.

It's important for the experts to reach consensus about the problem. When agencies fail to be firm, they show themselves as being weak and wishy-washy. They lose their credibility. Agencies need to step up to the plate and stick with their decisions. This is the right thing to do but they are afraid to side with industry. They turned control over to fringe environmental groups by not communicating with each other.

At the same time, it is difficult to achieve consensus when, as one key informant stated, 'we can't always characterize risk and it's difficult to communicate that to folks, especially if they are not equipped to deal with uncertainty.' When uncertainty leaves questions unanswered, community members begin wondering 'are they telling us everything? Are they hiding things? We might be breathing it and don't even know it.' Such a 'lack of trust makes discussions difficult.' In fact, one key informant suggested, 'probably things are going too well because people don't come to the meetings. They only come when they are distraught about something.' This suggests that the norm is for community members to participate only in specific issues that directly affect and/or upset them, and not to participate in general issues.

Factors that Decrease Participation in Wellpinit and Ford

One factor that discourages community members to participate in decision-making about DMC's Midnite Mine and mill site is the perception that environmental regulators and other outsiders do not understand 'what the community does,' and are 'not familiar with their lifestyle and culture.' 'Limited opportunities for interaction with community members' and low attendance at meetings, however, makes it difficult 'to get a handle on what their concerns are and if they want more local people involved.' From another perspective, a general lack of involvement in environmental issues stems from a poor understanding of how individual behavior may impact the environment. As one key informant explained,

> Part of this is cultural. It is hard to convince a person who burns plastic in his waste barrel and uses old oil or tires to start a fire in a brush pile that there is any thing to be concerned about. After all, I always dump my old oil on the ground and it hasn't hurt anything yet.

While key informants recognize 'there's people that don't care, there's only so much you can do,' the idea that participation will have little influence on decision-making is one cause of low involvement. For example, key informants reported that:

> The tribe dropped out because they thought it was a mockery of predisposed decision.

> I think that, aside from those with a financial interest in the operation and those that just don't care, it has left a number of folks feeling fairly impotent. They can go to a meeting and vent and yet it has no effect on how the company or the state do business.

> I was completely ignored. I made comments but was ignored and my questions were answered flippantly. The issues brought up were never given serious consideration.

Perceived 'divisions within the community' about what should be done with the mine and mill site leaves an impression that 'not many people off the reservation are interested outside of the federal agencies.' Slow progress with characterization and remediation activities is also a source of frustration. As key informants explained:

> This has been going on for years. People get frustrated and become apathetic. It's hard to see what's new.

> There's some community fatigue going on, sometimes people give up because the progress has been so slow.

> There are meetings about once a year and 25 to 30 people come. Part of it is because so little seems to be accomplished between meetings. There's nothing much for people to hang their hat on.

> Trying to get a community involved that's lived with the problem for so long is very difficult, we don't seem to even get the land owners involved.

Redirection away from the central issues to other matters and 'new people com[ing] in and want[ing] to start all over again' were other sources of delay. More specifically, one key informant reported that:

> The local 'complainers' were more interested in getting their names and faces in the news than they were in solving problems so at each meeting they would find a new face to bring up all of the same questions. I feel if they had spent half as much time trying to help we would have had a solution much sooner.

Another source of apathy reported was 'a feeling that you have to be a rocket scientist to understand all this stuff so people check out.' Since 'the technical jargon makes it hard to understand what's going on and hard to understand [how] the scientific process . . . affects the cultural process,' some people 'don't know how to get involved or what they should be concerned about.' Difficulties gaining access to technical information after making several requests for the same thing left one key informant wondering 'if information is passed on to the right ears. It's not consistent and sometimes it seems like things go into a black hole.' More specifically,

> There were times we didn't know what was coming down. We had to keep harping on them. I don't know if they were trying to hide things, sometimes they were. We had to constantly keep asking, kind of like what we didn't know wouldn't hurt us, or them. They wouldn't always give us water sample results. We had to keep pestering them and asking questions like where's the fence? Where's the samples? They just didn't want to tell us sometimes. We would find out things after the fact sometimes, like about the spills on the road.

In addition, 'not all of the public meetings were announced or advertised,' in a way that local people knew about them soon enough to be able to attend. Such circumstances build on the long established distrust between DMC and the community and serve as a source of friction in decision-making. Echoing the sentiment that 'the state and DMC are judged as the bad guys,' a key informant reported 'the state and DMC . . . were afraid they would say something wrong so they always had lawyers around.' In fact, 'some players have been involved in previous law suits against each other. It's been a source of distrust between the state, tribe and DMC.' The 'general lack of trust in the federal government' creates a dilemma for environmental regulators in that '[they] want to build trust and engage the community but don't know how.'

Factors that Increase Participation in Millersburg and Albany, Oregon

In spite of the fact that public participation in decision-making about TWCA is low, key informants suggested some things that encourage it. One action proposed to encourage participation in community-level decision-making was utilizing a third party to diffuse fear among participants. More specifically, one key informant suggested:

> [Public participation] works really well when we can get the company and the community in the same room but usually they are afraid of each other. It's much more effective when we can get them to talk to each other when we are the middle man.

One key informant reported that community members think 'regulatory agencies love public testimony, it's where people can spout out what they have to say and then they go away.' This makes a concerted effort to sincerely listen to participants important. As one public meeting participant wrote:

> People I have met who work at [TWCA] every day, who know first-hand what the situation is, are frustrated and concerned that their voices will not be heard. They fear that this 'public hearing' process is only a formality and that the EPA has already made its decision, and that decision will hurt their jobs (TWCA Administrative Record 14.4-0012640).

Hence, for the Millersburg and Albany community, the recognition of potential impacts to their employment base is critical. That is not to suggest one must guarantee long-term job stability, as that is not realistic, but one must be sensitive to their concerns about economic impacts since that is a fundamental and central issue for them. Another community member advised:

> As a long time Federal employee, I grieve over the low repute to which my old employer has fallen. I fear your contemplated action in this matter will further reinforce those attitudes of disrespect and distrust becoming so prevalent in the public mind (TWCA Administrative Record 14.4-0011817).

This illustrates how a little listening may go a long way.

Factors that Increase Participation in Wellpinit and Ford

Wellpinit and Ford, Washington, key informants also offered insights about factors that increase participation even though community involvement in public affairs is generally low. For example, 'people that are directly affected, are more interested' in participating in community-level decision-making. However, unclear and uncertain information may make it difficult for local residents to understand why their input is important. This requires environmental regulators to be both good risk communicators and good listeners. One key informant reported that 'the community is supportive of the clean-up but some don't feel they are doing enough and involving them enough, or addressing enough of their concerns.' Part of this concern stems from the well-rooted distrust community members have in environmental regulators and DMC. Minimizing distrust is likely to be the biggest challenge for the effective community-level decision-making about the DMC's

Midnite Mine and mill site. This will require asking, rather than telling the community how to involve them. Most importantly, as one key informant proposed, 'the company and the state need to listen to the community, they are the ones that live here.' The situation is not entirely hopeless. As one key informant explained,

> We can carry on a conversation with DMC now but for awhile we couldn't. The community was against them. They did listen to what we had to say in this particular instance. The company figured out it was better to work with us than against us.

This also suggests community members may be more likely to participate in decision-making activities if such activities focus on specific issues rather than vague generalities. Thus, breaking complex issues up into specific, manageable, pieces may be an effective strategy for improving community involvement.

Benefits of Participation in Millersburg and Albany

While some Millersburg and Albany residents may perceive public participation as nothing more than 'a necessary evil,' one reported benefit of public participation was that it allows 'people to feel they have control over their future.' Going a step further, one community member explained:

> As a citizen of Oregon I pride myself in being environmentally concerned and ecologically responsible. I feel I have the right to not only be informed about the processes at [TWCA] but to be protected by the federal government against any contaminants that result due to the industrial processes at this facility (TWCA Administrative Record 14.4-0000001).

Local residents also reported that public participation 'brought the community together about specific issues' and provided an opportunity to learn more about the problem. Because the preoccupation with job losses is so high in this community, however, community members are likely to remain somewhat reluctant to participate in decisions about TWCA on a large scale if doing so threatens the community's economic stability in some way. They are more likely to participate in issue specific matters that directly affect them.

Benefits of Participation in Wellpinit and Ford

Similarly, Wellpinit and Ford key informants reported that '[the public participation process] brought the community together, some of the people I hadn't seen in a long time came out. It put Ford on the map.' For one key informant, 'the most important thing to come from all of this is that the parent company is doing what should have been done by them from the very outset. Hooray!' Another key informant stated 'I learned a lot, I'm glad I was part of it but I'll be glad when it's done and over with.' Echoing that notion, another key informant said, 'it helped in regards that the public became educated, appointees already were. It helped getting the plan accepted . . . but it's very time consuming and expensive.' As previously discussed, distrust among environmental regulators, DMC and community members contributes to decision-making delays. However, this leaves the door open for improving the public participation process with an eye towards saving both time and money. Building upon perceived benefits of participation may make that even more possible.

Ways to Improve Participation in Millersburg and Albany

After nearly two decades of mitigation activities, one key informant stated that it was 'too soon to determine' how public participation could be improved. Other key informants, however, offered advice along three primary themes: 1) resources necessary for participation; 2) interaction between the public and environmental regulators; and 3) educational strategies. With regards to resources, one key informant called for more funds to facilitate public participation, pointing out that:

> For most people this isn't a full-time job, they have very limited time and money to put into these things. Most people don't have adequate resources to get very involved, they still have to work for a living. Resources need to be made available to public interveners [for communities] to make their case.

Another key informant proposed that large strides could be made by 'mostly respond[ing] to questions, by meeting and talking with [the community] . . . but there haven't really been any public meetings, there's no local interest in them.' An environmental regulator proposed:

> The biggest area for improvement is probably if the public could spend more time with us and we could spend more time with them. It goes both ways, to learn about our programs, how we regulate companies, the authority we have and the authority we don't have. Sometimes they get frustrated when we don't do things that we don't have the authority for . . . [the public could] ask us questions about how and what we do to regulate companies, ask us about company emissions. There's almost a paranoia about it. Someone might say 'I know late at night the company turns off their pollution control systems.' We have information that would tell us if that were the case.

Educational strategies suggested include 'step[ping] up public awareness about Superfund' in general. While recognizing that public participation 'would be very difficult to improve,' another key informant stated 'public participation is critical, it's essential to get information out to the public.' One place to begin would be 'distilling information in a way the average person can understand it and the notion of risk . . . so they could make their own decisions about what risks they are willing to accept.' As discussed in chapter five, technical jargon and explanations that lack connections to the daily lives of the Millersburg and Albany residents, build upon existing conflicts and distrust in environmental regulators. Hence, working with the community to develop meaningful risk explanations is critical for improving public participation in community-level decision-making, and relationships between environmental regulators and community members.

Ways to Improve Participation in Wellpinit and Ford

Many opportunities to improve public participation in community-level decision-making exist in Wellpinit and Ford as well. Indicating a need to educate community members about the risks present and potential remediation options in a meaningful way compatible with their culture, one key informant asked, 'how can housewives and farmers make decisions about such technical and complex issues?' One key informant suggested that 'to get at the root of the problem a person would have to understand the community well and that would take years.' Similarly, another key informant felt 'educating the community and regulatory agencies from both an environmental and human health effects perspective, as well as people involved with any other radioactive sites' was essential in order for people to make informed decisions. However, figuring out a meaningful way to communicate potential risks is not readily apparent. As one key informant asked, 'do we need workshops, a classroom setting forum, fact sheets, what's the best method?'

One way to begin answering that question would be 'to learn more about [the community] and to keep doing it [through perhaps] more regular, longer visits . . . more face-to-face communication.' But as environmental regulators admit, 'we're not sure who to target.' This is where working closely with the community to identify audiences and appropriate risk communication materials will be critical for improving public participation in community-level decision-making. With that in mind, one key informant suggested the process 'needs a coordinator tied to the community, with good communication skills, and no axes to grind, just objective technical assistance.' Other recommendations include 'maybe get some young blood in there, people that are going to be here longer,' 'getting the public involved,' and 'have meetings at lunch time so more people can attend them.' Most importantly, as one key informant stated:

> At times the science and the public are in conflict with each other but you can't throw one of them out. You're not going to get a solution that makes technical sense and is acceptable to the public that way. We must figure out how to work together.

Preliminary Conclusions

In spite of the differences between Millersburg and Albany, Oregon, and Wellpinit and Ford, Washington, these communities share similar challenges with respect to improving public participation in community-level decision making. Priorities for improving community involvement in both situations are getting community members to the decision-making table, and once they are there, establishing a meaningful risk communication process. For these efforts to be successful, the data presented here suggest environmental regulators and community members must work together to develop a strategy tailored to each community's culture. Since community members associated with both the TWCA and DMC cases are more likely to become involved in specific rather than general issues, breaking complex issues apart into specific, manageable pieces may increase community involvement. Considering that the health concerns of Millersburg and Albany, Oregon residents are greatest among TWCA workers, perhaps working with labor union representatives would be a useful avenue. In Wellpinit, Washington, working through the appropriate chains of command within the Spokane Tribe of Indians and employing verbal, rather than written, approaches will be important. Similarly, building upon relationships with Ford, Washington residents previously involved in decision-making about the DMC mill site provides a place to start.

The larger question is how does one facilitate public participation, especially given the mandated, top-down management strategies ingrained in environmental policies, established histories of low community involvement, and high levels of distrust in polluters as well as government officials which include environmental regulators? Moreover, the norms for some communities, urban and rural alike, may incorporate a minimal concern for environmental issues in general, coupled with marginal participation in community affairs. In response to these challenges, the data here indicate some specific underlying requirements that are necessary to meet in order to improve community involvement. First, it will be important to establish clear goals of how all parties will participate and commitment to those goals. These goals can serve as a yardstick for what community members, industrial entities and environmental regulators achieve as a group, providing an opportunity to build relationships based on accomplishments rather than failures and distrust.

Secondly, people need to have a clear role in community involvement activities. Participation goals will help make their roles and parameters of influence clearer. However, efforts to communicate how the risks may affect them and how they will benefit from participating in the community-level decision-making process are equally important. Components that risk communication strategies need to incorporate, as pointed out the key informants, include different perceptions and conflicting information about risks, the uncertainties of the potential risks, culturally sensitive language that acknowledges and respects the community's shared history and identity, and the recognition of pre-established distrust between environmental regulators, industrial entities and community members.

Thirdly, people need resources to participate. This may require financial assistance from the agencies overseeing mitigation decisions but will also require community support. This is a challenge that will not be easily met. Fourthly, opportunities for environmental regulators and community members to interact need to be established. Desires of all the parties involved need to be identified, including feasible times and locations for meetings. This too, can be part of participation goals. Clearly, this is only a beginning as proposed by the thoughts and insights of the key informants that participated in this project.

Finally, environmental regulators need to learn more about the communities they interact with. One approach to teach environmental regulators about communities may be working with local schools. For example, as a school project, students, teachers and parents could gather information about what makes their community important and how they define their community. Through already established techniques like public meetings and workshops, this information could be shared with environmental regulators and industrial entities. This creates an opportunity where communities can become part of the solution. Local educators may also be an excellent resource for developing culturally appropriate risk communication materials. Students could also review sample materials and provide insights about what risk information is unclear and how to make it more understandable. Perhaps the data presented here raise more questions than they answer, but that too, can be a point at which shared dialogues begin.

Chapter 8

Lessons Learned about Community Structure and Environmental Decision-making: Where Do We Go From Here?

Introduction

The focus of this project is to better understand the processes that environmental regulators, industrial entities and community members engage in as they define significant environmental risks and related mitigation actions, and how such processes impact community cohesion. The contribution of specific community characteristics to community-level decision-making examined in this project include: 1) shared history; 2) community identity (e.g., geographical boundaries, historical images, physical structures, stigma effects, and attachment to place); 3) control in local decisions; 4) distribution of power among local institutions; and 5) participation in decisions about environmental risks and mitigation. This chapter summarizes the roles of these characteristics and their interactions with respect to the operational premises presented in chapter two. Finally, recommendations for future study of community-level decision-making will be made. Increasing our understanding of these important social relationships may better equip both decision-makers and communities to effectively identify and manage environmental risks. This is especially important for the communities in this project as the types of risks targeted will be present in their landscapes indefinitely.

Shared History and Community Identity

Chapter four examined the degree to which community members share a common life and how their historical interactions formulate ideas about community identity, and in turn, environmental risks. It also considered how environmental regulators' understanding of shared history and community identity influence community-level decision-making processes. The first step, and perhaps the most important, in examining historical influences is to determine the appropriate starting point to ground the examination (White, 1991). While the point at which environmental contamination occurred or risks were detected may be a tempting beginning, this project illustrates much earlier time frames require consideration.

The communities of Millersburg and Albany, Oregon, for example, take pride in more than 150 years of individualism and status as the industrial hub for the Willamette Valley (Linn County Pioneer Memorial Association, 1979). As stated in chapter four, the building blocks of this industrial hub include: 1) favorable geographical conditions; 2) individual, predominantly white, land ownership where land was acquired by occupation largely in the absence of negotiation with its indigenous inhabitants; 3) transformation of indigenous land uses to place-fixed, long-term cultivation and manipulation of natural resources for economic gain and exchange; 4) marginal understanding and intolerance of indigenous traditions; and 5) insignificant resistance from indigenous groups to land transformations due to their reduction in numbers as a result of imported diseases. Today, industrial activities serve as a foundation for community members to establish and maintain relationships with each other, and securing long-term employment for community members remains the top community priority. Since Teledyne Wah Chang Albany is the area's largest employment this means, at least in part, maintaining TWCA as a fixture of the Millersburg and Albany community. In fact, a recent 7-month labor dispute at TWCA reminded community members of how dependent they are on TWCA economically. The strike also provided workers with an opportunity to strengthen their mutually dependent relationships rooted in union membership.

Generally speaking, Millersburg and Albany, Oregon residents perceive their long-standing and very dependent relationship with TWCA, a known polluter, to be positive. The pollution problems associated with TWCA, both past and present, tend to be perceived as both manageable and necessary for economic growth. Since the presence of TWCA is critical to the Millerburg and Albany community identity, many community members fear that excessive environmental regulations and mitigation may infer job losses and hence, community-wide economic demise. Conflicting perceptions about environmental risks held by outsiders are in many ways symbolically associated with their community identity, inadvertently turning debates about risks into attacks at their community identity. For example, concerns about pollution from TWCA operations is associated with being against TWCA in a general sense, to being anti-Millersburg and Albany, and anti-progress on the extreme end. These perceived attacks heighten distrust in regulators and encourage suspicion of mitigation activities. In fact, several community members suggested that they had more confidence and trust in TWCA than they do in environmental regulators.

This is not true for all community members, however. The recent labor dispute drew attention to long-standing worker health issues associated with TWCA that workers traditionally managed on an individual basis. Since the specific health effects from the mixture of contaminants present at TWCA is not fully understood, there is plenty of room to reframe issues once thought insignificant as new problems. Hence, the role TWCA plays in the Millersburg and Albany community identity is subject to change. The addition of new industries in the area has reframed TWCA's community role such that, as one key informant reported,

TWCA is 'not the only thing we run on anymore.' Just the same, the area's industrial focus makes the need for today's environmental regulations and mitigation difficult for many community members to fully understand and embrace.

Frustration with mitigation delays is not only associated with concerns about potential increases in unemployment but also with complex and confusing explanations of the risks present. This has left some community members wondering if mitigation is even necessary. The failure of environmental regulators to clearly answer community members questions, albeit questions entrenched in uncertainty, also exacerbates distrust in regulators and 'official' reports. Mitigation thus far expands across decades and the types of risks present require long-term management with no guaranteed end date. Hence, debates about risk management continue to threaten the Millerburg and Albany community identity. While environmental risks and mitigation have not significantly disrupted the social order of Millersburg and Albany, the potentially changing role of TWCA and changing ideas community members hold about TWCA may in the future. In the meantime, community members remain strongly attached to the Millersburg and Albany landscape, if only because that is where their jobs are. This attachment to place-based jobs may make it difficult for many community members to leave, even when leaving becomes a desirable option (Tolbert et al., 2002).

Wellpinit and Ford, Washington, residents on the other hand, incorporate interactive relationships with the land dating back to pre-colonization into their community identity. Community members tie relationships with the land to all activities. This, coupled with the fact that tribal members can practice traditional lifeways only on the reservation, reinforces their strong attachment to place. Permanent loss of land for non-tribal development is also part of that history and community identity. Moreover, the community members' process of defining community produces different geographical boundaries for their community than those utilized by environmental regulators. This is complicated by the fact that, as described in chapter six, uniform descriptions of the mine site among federal agencies and experts are lacking. Hence, when community members, industrial entities and environmental regulators sit down at the decision-making table, they bring with them different community frames. Failure to recognize these differences creates ample opportunity for conflicts about risks to evolve into symbolic conflicts about community identity and the loss of indigenous lands.

Wellpinit and Ford community members face many challenges on a daily basis including the struggle to maintain cultural traditions, subsistence-based lifestyles, and personal privacy. While the specifics of individual lifestyles vary, shared privacy desires and close relationships with the land function as the central components that establish a common life among community members. This in turn, may reinforce bonds among community members even when their interaction is minimal (Wilkinson, 1991; Bridger and Luloff, 1998). At the same time, community members recognize a need for economic development as well as how their landscape limits their development and mitigation options. They do not oppose all mining activities and have both positive, e.g., Sherwood Mine and Mill, and negative, e.g., Dawn Mining Company operations, experiences with uranium mining and milling operations, in particular.

These competing demands, however, bring the relative function of their land into question, and highlight differences in values among community members and differences in economic benefits rendered from DMC operations. Under these circumstances, mitigation provides a means to restore the lost culture community members associate with land use restrictions. The way community members frame mitigation may, however, differ from environmental regulators and technical experts. For example, technical experts tend to frame risks in terms of prediction and prevention whereas lay persons tend to frame risks in terms of detection and repair (Gray, 2003). Since the specific health effects associated with the mixture of contaminants present at the DMC Midnite Mine and Ford mill site are not explicitly clear, the need for and type of mitigation is open for debate. In combination with differences in how community is framed, different ideas about risks and mitigation actions may heighten underlying symbolic conflicts, especially if the goal of mitigation is to control risks rather than restore affected land.

In both the TWCA and DMC cases, the lack of recognition and understanding of the community's shared history and identity amplify existing conflicts in a number of ways. First of all, the failure to recognize differences in framing community creates an opportunity for outsiders to attack, albeit unintentionally, a community's identify. Secondly, the authoritative role of the lead federal agency to make decisions for locals that feel misunderstood may build upon existing distrust in environmental regulators. Thirdly, a less than complete understanding of how community members interact with their landscape may cause environmental regulators and technical experts to incorrectly estimate potential health risks. Similarly, different ideas about community frames and boundaries may improperly identify all affected parties. Both of these actions may further increase distrust in decision-makers as well as inhibit the ability to reach consensus about mitigation actions. The indefinite mitigation timelines in both the TWCA and DMC cases, and mitigation progress barely visible on an annual basis, also bread frustration and distrust in environmental regulators. The failure to recognize the extent to which distrust in federal agencies is incorporated into community identity may make matters even more contentious.

The need for environmental regulators to become knowledgeable about the community's shared history and identity is also critical for improving risk communication strategies in both situations. It will be equally important for each community to determine and communicate what forms and methods of information exchange best suit its members. As the data in this project point out, risk communication must become a shared responsibility and shared dialogue if it is to be successful. One way to begin acquiring knowledge about a community's shared history and identity would be to measure the variables suggested to be important in both the TWCA and DMC cases, including: 1) how community members ascribe boundaries to their community; 2) traditional community functions and roles; 3) community functions and roles imposed by others; 4) disruptions in community roles and functions; 5) meanings associated with community roles and functions;

6) attachment to place; 7) uniformity in ideas and values; and 8) uniformity in the distribution of potential health effects.

Control in Local Decisions and Distribution of Local Power

Chapters 5 and 6 examined how much local control the case communities had in decisions made about the Superfund sites studied as well as the distribution of that control across local institutions. As the data indicate, the role of TWCA as the Millersburg and Albany, Oregon, area's largest employer is not only important to the local economy, but also provides TWCA with the opportunity to dominate community-level decision-making. Community members appear to readily support such a position, sometimes denouncing the input of outsiders, particularly the Environmental Protection Agency. The recent labor dispute at TWCA, however, illustrates that community members who work for TWCA are able to organize issue-specific collective responses in opposition to TWCA. Hence, while the norm is for community members to generally accept and support TWCA's dominant role in local affairs, union membership provides a means to formulate collective, issue-specific, responses to TWCA's practices. This prevents TWCA from being entirely free of local scrutiny.

In addition, the mandated authority of the lead federal agency to make local decisions counters TWCA's dominant position. Lead agency authority may also privilege 'official' and technical risk information to the extent that it is assumed correct unless proven otherwise by community members (Brown, 1992; Swanson, 2001). Such challenges to traditional lines of authority by outsiders may contribute to underlying conflicts and encourage community members to discount external perspectives (Pfeffer et al., 2001). This also creates a situation where public meetings may deteriorate into defensive contests rather than exchanges of information. Furthermore, under these mandated circumstances, efforts to solicit community input are not only subject to extreme criticism, but even under the best conditions the lead agency's ultimate decision-making authority creates the perception that community members have little influence in local decisions. As the data suggest, it will not be easy to convince Millersburg and Albany community members that their perspective weighs in or that their participation in decision-making about TWCA is important.

Unlike TWCA, the Dawn Mining Company does not occupy a favorable, dominant position in the Wellpinit and Ford community. Instead, distrust in DMC, as well as federal agencies in general, is part of the Wellpinit and Ford community identity. This makes community members suspicious of DMC's actions and decisions made by outsiders about how to manage local lands. Moreover, personal actions are no longer private but rather the subject of outsider scrutiny. Since privacy and interactions with the landscape are principle tenements for a common life among Wellpinit and Ford community members, these conditions make it difficult for community members to retain their sense of place and maintain traditional lifeways. Additional land use constraints are likely to build upon underlying conflicts associated with historically imposed restrictions, leaving community members feeling even more powerless (Hanson, 2001; Lacy, 2000). Given their dependence on external networks to meet daily living needs, additional time constraints and limited resources may make it even more difficult for Wellpinit and Ford community members to influence local decisions in contrast to Millersburg and Albany community members. As a result, Wellpinit and Ford community members may readily become apathetic in response to the continuation of traditional paternalistic actions on behalf of federal agencies and see little reason to participate in community-level decision-making (Shrader-Frechette, 2002).

The Dawn Watch group, however, demonstrates that Wellpinit and Ford community members are capable of forming issue-specific collective responses, despite the absence of long-standing community organizations and limited collective resources. Limited resources may in fact, only allow for issue-specific collective responses (Sharp, 2001). Even under these constraints, such collective responses that promised to protect personal privacy and land use interests provided community members with an incentive to get involved in the New Jersey dirt import issue and to develop cohesive relationships that they may not have pursued otherwise. For Wellpinit and Ford community members, the New Jersey dirt issue turned out to be an empowering opportunity as was the TWCA labor dispute for TWCA workers. In both the TWCA and DMC cases, challenges to traditional community functions contributed to underlying conflicts and encouraged community members to discount external risk perspectives.

Despite the differences in shared history and community identity among the case communities, Millersburg and Albany, Oregon, and Wellpinit and Ford, Washington, community members encounter many of the same challenges when it comes to maintaining control in local decisions. In addition to a sense of powerless in local decisions overseen by lead federal regulatory agencies, the inability of the lead agency to answer questions clearly and the public airing of differences among federal and state agency personnel that occurred in both cases decreased community members' confidence and trust in said parties. Under these circumstances, community members are left wondering who is in charge and where to acquire information that empowers them to make informed personal decisions. Indirectly, little confidence and distrust in environmental regulators coupled with limited technical knowledge may heighten their fears about risks. An incomplete understanding of the area's geographical and social conditions on the part of environmental regulators further reduced confidence and trust in experts' assessments and decisions. Indefinite mitigation and post-mitigation monitoring time lines may also reinforce reduced confidence and trust in experts. To complicate matters further, other industries in the TWCA area prevent the reduction of risk altogether under the best circumstances. Likewise, for the Wellpinit and Ford community, naturally occurring sources of radon in the area allow for only risk reduction, not risk elimination. Thus, as much as community members may distrust environmental regulators, they find themselves in the awkward position of having to rely on outside expertise in new ways.

Under such conditions, establishing cohesive relationships between community members and environmental regulators is very challenging at best, and gridlock is very possible (Rosa and Clark, 1999; Williams, 2002). To minimize gridlock and conflict during decision-making processes, creating opportunities where communities maintain some sense of local control in mitigation decisions will be essential. This is especially important in situations where risk outcomes are unclear, frequently making the need and type of mitigation that is necessary and reasonable difficult to determine. Considering that technical experts tend to frame risks in terms of prediction and prevention whereas lay persons tend to frame risks in terms of detection and repair (Gray, 2003), the opportunity to associate debates about risks with attacks on community identity is great. In turn, threats to community identity may expand into debates about group legitimacy and encourage defensive behavior (Gray, 2003). These are hardly the ingredients of consensus. That is not to suggest that external agencies should cater to community desires without question or vice versa, but rather each needs to recognize the position of the other with open ears. To that end, the data suggests variables important in both the TWCA and DMC cases with respect to control in local decisions and distribution of power among local institutions include: 1) general dependency on external ties; 2) need for external funds to mitigate environmental risks; 3) quality of past relationships with regulatory agencies; 4) location of the parties responsible for environmental risks within the power structure of the community; 5) location and role of technical experts within the power structure of the community; 6) identification of and responses to victims; 7) community members' ability to organize issue-specific collective responses; and 8) degree of uncertainty associated with environmental risks and mitigation actions.

Participation in Decisions about Environmental Risks and Mitigation

Chapter seven examined how community members and environmental regulators participate in decisions about the TWCA and Dawn DMC's Midnite Mine and mill site. In communities associated with both the TWCA and DMC cases, factors that amplify conflicts and risk perceptions, and hinder participation in community-level decision-making include: 1) incomplete, unclear and uncertain risk information; 2) unclear or limited roles in decision-making processes; 3) disrespect and poor understanding of local community history and identity; 4) unclear risk characterization and risk management goals; 5) distrust in environmental regulators as well as parties potentially responsible for the environmental risks present; 6) personal time constraints; and 7) lack of resources to participate. In both the TWCA and DMC cases, community members were more likely to participate in specific, rather than general, community issues. Millersburg and Albany, Oregon, residents may resist participating in activities that oppose TWCA in general, as TWCA is an import fixture in their community and attacking TWCA may mean attacking their own community identity. The desirability of DMC's presence in the

Wellpinit and Ford community, however, has been mixed historically such that the removal of DMC may strengthen, rather than harm, their community identity.

Key informants also suggested ingredients for improving risk communication strategies. First, breaking complex issues down into specific, manageable pieces may increase community involvement. Establishing clear roles for participation and goals of participation in decision making processes are important for encouraging community involvement, as well as providing resources for community members to participate. In addition, finding ways for community members and environmental regulators to interact may help build less adversarial relationships. Environmental regulators also need to make a concerted effort to learn about the communities that they interact with. Components that risk communication strategies need to incorporate, as pointed out the key informants, include different perceptions and conflicting information about risks, the uncertainties of the potential risks, culturally sensitive language that acknowledges and respects the community's shared history and identity, and the recognition of pre-established distrust between environmental regulators and community members. In terms of measuring variables to identify strategies to both improve community involvement and better understand barriers that may inhibit participation, those suggested by the data from this project include: 1) lines of authority; 2) opportunities to participate in issue-specific matters; 3) opportunities to participate in general community matters; 5) community members' roles in participation; 5) history of conflict and cooperation between community members, industrial entities and environmental regulators; 6) benefits associated with participation; 7) community norms concerning participation; and 8) community norms about the risks in question. Increasing our understanding of these important social relationships may better equip both decision-makers and communities to effectively manage environmental risks.

An Evaluation of the Operational Premises

Operational Premise 1: The more uniform the ideas environmental regulators, industrial entities and community members hold about affected parties, the more likely it is that they will reach consensus about the significance of environmental risks and need for mitigation.

The data suggest that in both cases, community members, industrial entities and environmental regulators frame community in different ways. Millersburg and Albany, Oregon, community members begin their history with roughly 150 years as an industrial hub in the Willamette Valley. Industrial entities are an important and valued part of that history, especially TWCA, the area's largest employer. Wellpinit and Ford, Washington community members incorporate relationships with the landscape preceding colonization into their community identity, as well as the mixed blessings associated with industrial entities. Environmental regulators, however, focus on the physical parameters of contaminants and this results in a cursory understanding of community social features. These differences in community frames also serve as a source of underlying conflict at the decision-making table.

Consensus about the health effects associated with the TWCA site, and DMC's Midnite Mine and mill site is not evident for a number of reasons. First, little is known about the specific health and ecological outcomes associated with the mixtures of contaminants present at the TWCA and DMC facilities. As discussed in detail in chapters five and six, the risk data available for both cases consists of extrapolations from animal data and estimated probabilities based on a less than perfect understanding of how community members interact with their landscapes. Others argue cause for concern is not necessary as long as no humans and animals come in contact with the substances as governed by administrative controls. With the presence of several other industries in TWCA's immediate area, and naturally occurring radon sources widely spread throughout the DMC area, it is impossible to eliminate risk completely, even given the best scenario. There is some shared concern about increased risks among workers in both situations but much less regarding community-level risks. More specifically, debates over what the risks are, who is most affected, and the need for remedial action continues for both cases, even after nearly two decades of mitigation activities at TWCA.

In the communities associated with both TWCA and DMC, lack of recognition of the different community frames that environmental regulators, industrial entities and community members utilize contributes to conflict, as does uncertainties associated with the hazards present. Being insensitive to distrust among environmental regulators, industrial entities and community members further encourages contentious mitigation discussions. An incomplete understanding of the affected community also reduces confidence and trust in experts' assessments. This makes the need for environmental regulators to learn more about the social dynamics of affected communities import for minimizing conflict as well as for properly identifying risks, especially given their lead authority role. More specifically, the data suggest it is essential to involve community members as soon as possible, particularly with respect to site assessment activities in order to develop likely exposure scenarios, rather than waiting to solicit community input at public meetings concerning mitigation plans. Given the subsistence-based lifestyle of Wellpinit and Ford community members, such efforts are crucial. Providing timely access to information about suspected risks and mitigation activities in a common language are important risk communication goals for both cases, especially since mitigation endeavors are long-term, and will also be important for reducing conflict.

Operational Premise 2: The more uniform the ideas environmental regulators, industrial entities and community members hold about affected parties and public health risks, the more likely it is that the affected community will cohesively support mitigation decisions.

Again, ideas about community frames and public health risks are not uniform in either the TWCA or DMC cases. Slow progress and ingrained distrust feed frustration among community members at both sites. Support for mitigation of TWCA historically and currently competes with economic interests to retain TWCA jobs. This makes Millersburg and Albany community members and industrial entities raise concerns about the need for mitigation. At the same time, TWCA workers are not free from occupational safety issues. For Wellpinit and Ford, restoration activities are welcomed but even under the best circumstances, will most likely never return affected lands to pre-mining conditions. Controlling risks—perhaps the only realistic option—is likely to be a disappointment to many as it implies a permanent loss of access to traditional hunting and gathering grounds, and in turn, a loss of culture for the Spokane Tribe of Indians. In both situations, it will be important for environmental regulators to work closely with the affected community to formulate risk communication strategies that are culturally compatible. Components that risk communication strategies need to incorporate, as pointed out by the key informants, include different perceptions and conflicting information about risks, the uncertainties of the potential risks, culturally sensitive language that acknowledges and respects the community's shared history and identity, and the recognition of pre-established distrust between environmental regulators, industrial entities and community members.

Operational Premise 3: A decrease in consensus among environmental regulators, industrial entities and community members will produce an increase in cohesion among community subgroups organized along lines of conflict, and a decrease in community-wide cohesion.

Those that expressed concerns about increased cancer in relationship to TWCA employment, generally interpret risk reduction largely as a personal responsibility that comes with the job and engage in individual responses, e.g., bringing bottled water to work, rather than collective responses. However, a recent labor dispute demonstrated TWCA workers' ability to formulate issue-specific collective responses. One of the issues under negotiation during the strike was maintaining health care benefits for retired workers. To that end, the strike provided an opportunity to develop cohesive relationships within subgroups formed as a result of common and very specific interests. The close physical proximity and utilization of common locations offer support to maintain such cohesive relationships. While there is no clear mass exodus from the area as a result of health or ecological concerns related to the TWCA site, one should not infer all community members have become at ease with the risks present just the same.

With respect to the DMC case, the Dawn Watch group is a good example of how Wellpinit and Ford, Washington, community members coordinate efforts around commonly shared concerns. Dawn Watch was a coming together of tribal and nontribal community members for a shared and very specific cause—to prevent the import of the New Jersey dirt. Community members also successfully acquired Superfund resources for remediation of the Midnite Mine. Once accomplished, many people returned to focussing on their individual needs and maintaining personal privacy. While there is little evidence of long-standing community-wide organizations outside of the tribe, Wellpinit and Ford, Washington residents are not without power but rather, selectively exercise and organize their efforts for specific interests in spite of personal differences. For them, specific interests build cohesive relationships rather that reduce cohesion among community members. Hence, as both cases suggest community-wide, generally cohesive relationships among community members is not a prerequisite for forming collective, issue-specific responses to environmental risks. Likewise, cohesive relationships within subgroups did not infer reduced overall community cohesion in either of the cases.

Where Do We Go From Here?

The data support the consideration of shared history, community identity, local control in local decisions, distribution of power among local institutions, and participation as important features in community-level decision-making about environmental risks and mitigation. Incorporating local knowledge into decision-making processes will improve decision-makers' efforts to properly identify affected parties and the magnitude of effects as well as underlying conflicts that may deter agreement about mitigation actions. In turn, such efforts will lay the groundwork for making comparisons between other cases, providing opportunities to learn from both decision-making successes and failures. After all, understanding risk is not just about understanding contaminants, but also about understanding how people interact with their landscapes.

As discussed in detail in chapters 5 and 6 for example, community members and environmental regulators associated with both the TWCA and DMC cases made comparisons to Love Canal, the site that founded the Superfund program. In contrast to Love Canal, none of the communities in this study experienced significant disruptions in daily routines or residential relocations even though the hazard ranking scores for the sites under investigation were similar to Love Canal (Midnite Mine = 50; Love Canal = 54; TWCA = 54.27). Since it is impossible for communities receiving federal mitigation funds to remain anonymous, the potential for stigma associated with hazardous wastes is likely and thus, important to consider but it is not clear how environmental stigma may impact the communities studied. In both the TWCA and DMC cases, attachment to place, albeit for different reasons, discouraged community members from leaving their communities in significant numbers. The scope of this project, however, does not extend to an examination of who may not be moving into the affected communities, and why. For the communities associated with TWCA and DMC, the recruitment of new industries has been a long-standing challenge but key informants did not indicate business recruitment has become more challenging following Superfund status designation. One TWCA key informant and two DMC key informants raised concerns about stigma associated with mitigation delays potentially hampering future business ventures so this may be an issue worth exploring down the road.

The necessity of community cohesion for effective community-level decision-making is not clear in this project and thus, is a topic warranting further investigation. Both the Millersburg-Albany community and Wellpinit-Ford community demonstrated the ability to organize collective responses to specific concerns even though community members generally focus on meeting individual needs rather than group needs. The data also suggests marginal involvement in community affairs may be the norm for the communities studied. At the same time, the data demonstrates that it is not wise for external organizations to underestimate the ability of community members, even in discordant communities, to organize collective responses—especially around specific issues. For example, the rural character and preoccupation with personal privacy among Wellpinit and Ford, Washington, residents did not deter their efforts to prevent the importation of the New Jersey dirt, in spite of their dependency on external resources for remediation of the Midnite Mine. Furthermore, the critical role that TWCA plays in the community identity of Millersburg and Albany did not prevent TWCA workers from entering a 7-month labor dispute.

Another important consideration for future research is mitigation time lines. While mitigation is nearly complete at TWCA, the long-term management of the risks present does not come with a guaranteed end date. The types of contaminants involved, the potential for a failure in current containment measures and/or administrative controls, and the potential discovery of new problems, leaves the mitigation door open indefinitely. The ability of everyone to remember the location of the hazards as new industrial development takes place poses another challenge, further complicated by the little confidence local people have in environmental regulators. Similarly, the Midnite Mine water treatment facility costs roughly $1 million per year to operate with no termination date planned at this time. Experts specified that the mine and mill site areas must be monitored for 200 to 1,000 years if current conditions remain stable (Washington State Department of Health, 1991). This raises additional concerns about resources for long-term risk management as well as the ability of everyone to remember the location of the hazards, especially given the indigenous trees and native grasses planted on top of them, not to mention the little confidence local people have in the environmental regulators.

Clearly, the challenges that lie ahead for the communities associated with the TWCA and DMC sites are not short in number. Differences in ideas about risks and land use between community members and environmental regulators may contribute to contentious decision-making. That being the case, it will be important for those involved in community-level decision-making processes to develop realistic expectations about erasing ingrained distrust and transforming top-down risk management approaches overnight so as to not worsen the situation. Specific recommendations for improving community involvement in decision-making follow.

Recommendations for Improving Community-level Decision-making about Environmental Risks

1. Begin interactions between environmental regulators and community members with a meeting where environmental regulators focus on learning about the community from community members. Coordinating such efforts through local schools and classroom projects may be a useful way to utilize resources.
2. Ask community members about previous relationships with environmental regulators, what worked well and what did not work well. This may help identity avenues to minimize distrust.
3. Ask community members what forms of communication are most useful to them. The key informant interviews for this study took place in a variety of forms to accommodate individual needs and schedules such as telephone and in-person conversations. Submitting comments in written form including email works best for others. Electronic forums in conjunction with public meetings may be useful for some communities. It may also be necessary to develop a mechanism to protect the confidentiality of community members so that they feel more comfortable sharing information with environmental regulators.
4. Provide opportunities for community members to be part of the solution. In addition to giving community members the opportunity to educate environmental regulators about their community, it may be useful to involve community members in conducting interviews with their peers to gather information about risks and how people interact with local landscapes. The use of temporary employees and volunteers has proven to be a successful way to gather census information in hard to reach neighborhoods. Such a strategy may be useful for gathering background social information relevant to risk identification in Superfund communities as well.
5. If possible, involve community representatives in the selection of facilitators for meetings. Perhaps a respected community member may effectively serve as a facilitator.
6. Discuss time lines for assessment and mitigation actions with community members so they can develop realistic expectations. Also ask for insights about how time lines might be shortened. This provides another opportunity for community members to be part of the solution.
7. Ask community members what they want to participate in. They may only want to participate in specific types of activities concerning specific issues.
8. Establish rules for group interaction to minimize the potential of risk debates being connected to underlying symbolic conflicts.
9. Most importantly, listen, listen, listen!

Final Thoughts

The issues addressed in this project are not simple nor are these communities the only ones struggling with them. But these cases are opportunities we can learn from. Most importantly, the approach utilized in this project demonstrates how the incorporation of sociological and biological knowledge can improve current strategies to identify and manage environmental risks. To that end, the *social amplification of risk* framework provided a means to incorporate psychological, social and cultural processes involved in community-level decision-making about environmental risks. This framework proposes risk messages pass through a variety of filters such as personal experiences, confidence in institutions, alienation from community affairs and perceived fairness of risk management processes (Kasperson et al., 1988). While this framework recognizes elaborate descriptions may be distracting, it is important to note that one has to not only receive but also understand risk messages before translation processes come into play. The extent to which community members' participate in community-level decision-making about environmental risks is in part limited by the availability of information and their skills to interpret technical information, even when such information is free of ambiguity. The point here is that psychological, social and cultural processes do not only filter risk messages but they may also block the transmission of messages altogether. Given the top down management approach ingrained in the Superfund process and lead federal agency authority to make local decisions, there was ample room to improve both the availability and clarity of risk messages to community members associated with the cases studied. Hence, improving an understanding of risk information will be dependent on increased community involvement in risk communication strategies.

The top down management approach and lead agency authority in the Superfund process also alter the direction of rippling impacts proposed by the *social amplification of risk* framework. This framework suggests signal filtering begins with individual processing, rippling outward to organizations, communities, and so on. In the case of Superfund, however, rippling impacts largely take place in the opposite direction. While individuals may be aware of hazards present in their landscapes, it is the lead federal agency that determines the significance of those hazards and that has the ultimate authority to determine how to mitigate those hazards. Processes internal to the lead federal agency filter what information becomes available to individuals. As discussed earlier, this technical information is generally assumed correct until proven otherwise (Brown, 1992; Swanson, 2001). This places community members in a situation where they may have to rely on technical experts in new ways. In order to make sense of such technical information, community members may also find themselves drawing upon personal experiences, information from family, friends, and coworkers, as well as other community and external sources. Hence, when trying to understand community-level decision-making processes, one must consider rippling impacts that begin at the lead agency level, moving inward to the individual level, and then outwards to personal, community and cultural resources. This also makes the need to understand how community structures filter risk messages all the more important in order to improve risk management strategies.

Using the *social amplification of risk* framework in conjunction with community theory, social constructionism and disaster research in order to better understand interactions between biophysical and social features, specific community-level risk amplifiers identified here include: 1) exclusion of shared history and community identity from efforts to determine routes of potential exposures to risks; 2) challenges to traditional community functions and lines of authority; 3) differences in how community members and environmental regulators frame potentially affected communities; 4) marginal and/or uncertain technical knowledge; 5) distrust and little confidence in environmental regulators and/or potentially responsible parties; 6) dependency on external ties to meet daily needs and fund mitigation; 7) limited resources to participate in technical decisions and formulate collective responses to environmental risks; and 8) unclear roles and goals of community-level participation. Not only do these factors function as risk amplifiers, they also amplify underlying conflicts at the decision-making table. Identifying and better understanding factors that amplify community-level risk perceptions and related responses will not only allow us to develop more informed policies, but also provides opportunities to improve community involvement in mitigation efforts. As the sorts of wastes targeted in this study will require our management for many years to come, so too, will these important social issues. May the learning, not the wastes alone, guide our way.

Appendix A

Key Informant Interview Questions

Key Informant Interview Questions

1. What interests do you represent?
2. How were you selected to represent these interests?
3. Tell me about the history of the site. What are the most significant problems, concerns, challenges?
4. How would you describe the population affected by the site and related decisions?
   [Who are they? What are their values? What is their culture like? How important are symbols and structures (e.g., buildings) for defining the community? What role does industry and industrial potentially responsible parties associated with environmental threats play in community identity?]
5. How have environmental threats impacted community activities?
6. Have environmental threats stigmatized the community? If, so how?
7. How have you/your group participated in decision-making about the site? How long have you/your group been involved in decision-making processes about the site?
8. How much time/money/resources has your group put into issues regarding the site?
9. How would you describe the distribution of power among decision-making representatives?
10. What barriers to participation have you/your group experienced?
11. What concerns are not represented? Over-represented?
12. Does public participation and consultation work? How so? How can it be improved?

Appendix B

Detailed Discussion of Pre-Columbus Culture and Congressional Acts Promoting Colonization of the Oregon Territory

Pre-Columbus Culture

The first indigenous peoples of the Willamette Valley of Oregon and the Spokane Valley of Washington persisted for as long as 10,000 years before European encounters. These indigenous residents were generally a peaceful people that depended on natural resources within their domains for survival. Working together in groups, they harvested or gathered what they needed from the land (Johnson, 1904; Scott, 1924). Not only was their relationship to the land the heart of their culture and livelihood, but the features of the landscape molded their culture (Clark, 1927; Mackey, 1974). A more specific examination of indigenous interactions with the Willamette and Spokane Valley landscapes follows.

Indigenous People of the Willamette Valley

There were two primary indigenous groups in the Willamette (meaning to spill or pour water) Valley of Oregon, the Upper Chinookians and Kalapuyans (Mackey, 1974). The Upper Chinookians consisted of the Wapatoos and the Cloughwallahs. The Wapatoo Indians lived at the mouth of the Willamette River where it enters the Columbia River. The Cloughwallahs, also known as the Willamettes, lived near the Willamette Falls, south of the Columbia River by present-day Oregon City (Clark, 1927). Unique to the Chinook culture was the practice of head binding. While a baby was sleeping, the mother would strap the baby to a board, covered with moss, and bind the infant's head with a leather band over a smooth piece of cedar bark. This caused the soft bones of the head to flatten, and after about one year the resulting depressions served as a mark of freedom. For girls, the more extreme the depression, the higher the dowry would be at time of marriage (Bowen, 1978; Clark, 1927; Johnson , 1904).

The Kalapuyans lived in the Willamette Valley south of the Willamette Falls (Mackey, 1974). Rather than uniting under one chief, each Kalapuya band had its own headman. All of the Kalapuya spoke the Calapooya dialect, a division within the Kalapooian linguistic stock (Mackey, 1974). Given this lingustic lineage, some call this same, 'most distinctly Oregonian' but least known, indigenous group the Calapooians (Clark, 1927, p. 42). To make matters even more confusing, observers and anthropologists link as many as 19 different names to the various bands of the Kalapuyans. Some of the confusion about different names for the Willamette Valley bands stems from the fact that white settlers named groups based on the geographic fixtures they lived by. For example, multiple sources refer to a band of the Kalapuyans living along the Calapooya River extending between present-day Albany and Brownsville in Linn County, as the Calapooya. Others ascribed names to the bands that fit within the Calapooya dialect (Clark, 1927; Mackey, 1974).

The Chinooks and Kalapuyans did not reside near large game grounds or use horses and thus, relied on small animal hides (muskrats, wood rat, geese, etc.) and cedar bark for clothing. Clay paint, feathers, bear claws, animal teeth and shells provided additional decoration. In the summer, the men would frequently go naked. Women spent their days as carriers, cooks, food gatherers, mat, hat and basket makers. The men hunted, made weapons and war when necessary. Both tribes fashioned utensils, pots, dishes and baskets from wood. The Kalapuyans crafted particularly exquisite baskets, so tightly woven that they were capable of holding water. The Chinook constructed fine, sea-worthy canoes capable of carrying up to as many as 30 persons. Occasionally, canoes aided maritime war efforts but the under-armed tribes generally did not pose a significant threat (Bowen, 1978; Clark, 1927; Johnson, 1904). Both the Chinooks and Kalapuyans controlled population size by practicing infanticide, especially in times of significant poverty. Both groups also buried their dead with their earthly possessions (Clark, 1927).

The Willamette Valley tribes had their differences, however. The Chinookian diet primarily consisted of fish caught by using poles and nets, then dried. The Kalapuyans on the other hand, relied upon a vegetarian menu of dried and roasted roots like Camas and wappato, supplemented with seeds, fruits, and berries gathered from nearly 50 plants. Temporary shelters made of poles and mats provided housing compatible with the Kalapuyans' food gathering. The Chinooks built plank houses over two to three feet excavations, extending as long as 200 to 300 feet, then lined and subdivided with mats. While very amenable to winter conditions, sewage disposal was problematic (Clark, 1927; Johnson, 1904; Mackey, 1974). In contrast to the soft and harmonious Calapooia language, the Chinook dialect contained distinct, guttural sounds. Since the Chinook engaged in trade to supplement their lifestyle, this frequently made transactions difficult. Items exchanged for fish included guns, bullets, beads, blankets, knives, hatchets, tobacco, metal pots and bracelets, and slaves. Rarely did the Kalapuyans trade or use slaves as they had few articles of interest to others to trade (Clark, 1927; Johnson, 1904).

According to two key informants, some of the people who live in the Willamette Valley refer to its indigenous inhabitants along the Calapooya River between Albany and Brownsville as the 'Mound Builders.' Leveled by centuries of storms, 151 remaining mounds stand three to ten feet high, and measure from 50 to 100 feet in diameter (Clark, 1927). While there are now laws to prevent bulldozing of the mounds, it is not clear how many there once were. Oak trees as old as 300 years top some of the larger mounds (Mackey, 1974). Items found in various mounds include bodies and ashes of the dead, arrow heads, hollow copper tubes, sun dials, stone axes, cooking bowls, spear heads, other stone implements and personal belongings (Clark, 1927; Mackey, 1974). Some believe the indigenous people that existed before the Calapooya created the mounds, others believe the Calapooya created them, but according to the two key informants, all agree the mounds are an unraveled mystery.

Indigenous People of the Spokane Valley

Aboriginal lands inhabited for centuries by the Spokanes (meaning children of the sun), an Interior Salish group, encompassed approximately three million acres in northeastern Washington, northern Idaho and western Montana. Historically, each of the three bands within the Spokane tribe (upper, middle and lower) had their own chief. The lands of the Spokane Valley affected both the social and economic life of the Spokane. In the spring, small groups dispersed from winter camps to gather food, hunt and fish. Activities of summer and early fall included root digging, and berry picking. This was also a time for many intertribal social gatherings, now known as 'Pow Wows.' In the late fall and early winter, smaller units would regroup along rivers and creeks that provided access to water and shelter during the winter months. The winter months were an important time for trading, visiting and observing ceremonies. Salmon was a principal component of their diet and a critical commodity for trading (Wynecoop, 1969).

In early times, the Spokanes supplemented their supplies by raiding other tribes. Raiding parties sought horses, food, weapons, and women for slaves. Unlike neighboring tribes, the Spokanes were typically not warlike people. They frequently formed alliances with other tribes including the Kalispels, Flatheads, and Coeur d' Alenes. This was a particularly useful arrangement for buffalo hunting and trading expeditions. Women wore long dresses, leggings, a belt and moccasins fashioned from buffalo, elk and deer hides. Men wore buckskin shirts, leggings, a belt, breech cloth, moccasins and sometimes, fur hats. Shelters consisted of teepee frames covered with mats of tule and rectangular housing at permanent camps to accommodate larger gatherings and ceremonies. Women made mats, bags, baskets, clothing, dressed the hides, gathered fuel, dug roots and prepared meals. Men made ceremonial clothing and weapons, hunted, cared for the horses, and made war when necessary (Wynecoop, 1969).

Summary

In summary, for centuries the indigenous inhabitants of both the Spokane and Willamette valleys maintained reciprocal relationships with the land such that the land shaped their culture and their culture shaped the land. In this sense, the loss of land would cause not only physical hardship, but would also change how indigenous peoples maintain traditional ways of life. With these changes came a loss of culture. This makes mitigation decisions not just about the loss of land deemed unsafe to human health but also about the loss of culture. New land use restrictions may build upon historical losses. Thus, understanding the effects of colonization on indigenous groups is important if we are to uncover underlying sources of conflict that may contribute to contentious decision-making.

Colonization

Colonization, the settlement and development of foreign territories by non-indigenous peoples, changed how the indigenous peoples of the Willamette and Spokane valleys interacted with their physical environments. Through a number of economic development activities and legal arrangements, land ownership shifted from indigenous inhabitants to white settlers (White, 1991). This shift introduced new relationships with the land. Stationary lifestyles and individual land ownership replaced seasonally nomadic lifestyles within a communal land system. The new land owners viewed natural resources as assets subject to manipulation for financial gain and personal betterment. This strongly contrasted the indigenous peoples' belief that natural resources are both vehicles for survival and cultural assets (Alley, 1889; Clark, 1927; Johnson, 1904; Wynecoop, 1969). Specific mechanisms of colonization in the Willamette and Spokane valleys include trading, missionary work, several laws and congressional acts encouraging white settlers to occupy the West, the removal of indigenous groups to reservations, and the expansion of the railroad. Examining the impact of colonization will help decision-makers working with Native American populations become aware of potential underlying conflicts that may deter successful mitigation.

First Encounters with Whites and the Establishment of Trading

The first encounters with indigenous people of the Pacific Northwest began with a Spanish expedition. Departing from California and heading north up the coastline, the expedition arrived in Nootka Sound on present-day Vancouver Island, British Columbia, in 1774. The Spanish laid claim to the Pacific Northwest territory on a second voyage in 1775. In 1778, the English Captain James Cook and his two ships reached Nootka Sound. The crew acquired 1,500 otter pelts from local tribes to use as clothing on their return voyage. Instead, they sold the pelts in China for a handsome profit. The word of potential fortunes in the Pacific Northwest quickly spread, bringing numerous English and American traders to the area (Clark, 1928; Johnson, 1904).

The influx of English and American traders alarmed the Spanish. Growing tensions brought Spain and England to the brink of war in 1790. In efforts to resolve the dispute, Spain and England signed the Nootka Convention in which both parties agreed to restore or compensate each other for any lands seized. On

23 March 1795, Spain vacated the Port of Nootka and terminated further interests in occupying the Pacific Northwest. Britain and America continued to jointly occupy the area (Clark, 1928; Johnson, 1904). Disregarding the fact that indigenous peoples inhabited the area long before their arrival, the British and Americans failed to recognize the unsubstantiated nature of their land claims.

With an eye on expanding both trade and personal fortune, the American fur trader Robert Gray sought out additional water passages in the Puget Sound in 1792. He sailed a river he named after his ship, the 'Columbia,' upstream about 25 miles only to ground on a sandbar (Clark, 1928; Johnson, 1904). Once the ship was free, he returned to the ocean. Shortly after word of Gray's discovery, British Captain Vancouver and two consorts attempted to explore the Columbia River but only one of the ships, the 'Chatham,' successfully navigated the sandbar. Vancouver concluded that the river had insufficient water for large vessels. After anchoring the 'Chatham' in a nearby bay just past the sandbar, Commander Borughton explored some one hundred and twenty miles upstream in a smaller boat manned by a crew of oarsmen. He ended his upstream journey when he reached the mouth of the Willamette River, naming the point Fort Vancouver. On this venture, Borughton and his crew were the first white men on record to encounter the Willamette tribes (Clark, 1928; Johnson, 1904). Some 21 years later, the North West Company established the first trading post in the area. The site of this 1813 post was on the banks of the Columbia River, opposite the mouth of the Willamette River (Clark, 1928). In 1814 the Hudson Bay Company built a trading post nearby. The two companies merged in 1821 with the Hudson Bay Company formally taking over in 1824. The following year, the Hudson Bay Company built Fort Vancouver (Clark, 1928; Johnson, 1904).

While trade routes to the Willamette Valley became more readily known in part due to a more manageable terrain, others sought an inland passage from the plains to the Pacific Ocean. Lewis and Clark were the first to explore the Missouri and Columbia rivers from 1804 to 1806 (Johnson, 1904). Tales of their travels encouraged other fur traders to explore the potential markets. In 1807, one such hopeful Canadian trapper, David Thompson of the Northwest Fur Trading Company, encountered the Spokanes and became the first white on record to enter Spokane territory. Three years later, Thompson established a trading post, referred to as the Spokane House, at the confluence of the Little Spokane and Spokane rivers. Tobacco and guns were popular items to trade for furs. The usual exchange was 20 pelts for one gun, a very lucrative market for the whites since one pelt was almost equivalent to one gun on the European market. In favor of a more convenient location on the Columbia River fur trade route, the Northwest Fur Trading Company built a new post on the Upper Columbia River in 1825. As the new post, Fort Colville, became the area's most important trading center, use of the Spokane House waned. In 1826, the Spokane House permanently closed (Wynecoop, 1969). All the while, Americans and British jointly occupied the area, questioning each other's territorial claims. However, they did not make any agreements about hunting and land use with the indigenous tribes.

Early Missionaries

As the fur trade established posts throughout the Pacific Northwest, a New England school teacher, Hall Jackson Kelley, began recruiting settlers in 1817 to spread Christianity via organized colonization (Johnson, 1904). His underlying agenda was to wrestle land away from the English and expand the American territory through occupation by squatters. In 1829, Kelley founded 'The American Society for the Settlement of the Oregon Territory' (Johnson, 1904, p. 213). Although Congress did not approve of Kelley's colonization scheme, Kelley continued his efforts. He undertook a number of expeditions, none of which were particularly successful. Even as his health failed, he hoped to inspire others to follow his lead and continued to write about the glories of Oregon (Clark, 1927; Johnson, 1904).

Unintentionally aiding Kelley's cause, four Flathead and Nez Percé tribal members arrived in St. Louis, Missouri, in 1831, purportedly sent by their tribes to acquire 'the book which would teach the red man more of the white man's God' (Johnson, 1904, p. 194). While in St. Louis, two members in their party died, and the other two, leaving empty handed, died before reaching their home villages. Their disastrous quest, however, received considerable publicity (White, 1991). In response, the Methodists were eager to begin missionary work among the Pacific Northwest tribes but were not sure where to start. Dr. McLoughlin at the Hudson Bay Company's Fort Vancouver post persuaded them to focus their efforts on the Willamette Valley. Subsequently, Methodist Reverend Jason Lee founded the first mission in the Willamette Valley in 1834. Four years later, two Catholic priests, Blanchet and Demers, began missionary work at Fort Vancouver. Both faiths competed for the Native Americans' attention. Each tried to claim superiority and prove the other false. This religious feud, coupled with little conversion success largely due to the missionaries' intolerance of indigenous traditions, led to the abandonment of the Methodist mission in 1844 (Johnson, 1904; White, 1991). Clearly indicating intolerance for the Native Americans, A.F. Parker suggested the failure of the missions stemmed from the inherent nature of the Native Americans, rather than from a lack of effort on the missionaries' part to understand the indigenous culture. In his appraisal of the situation he stated:

> Practically, the Indian is the same blanketed, unwashed, unenlightened savage he was when the peace policy was inaugurated. . . . But it was found impossible to Christianize the Indian. His mind could not or would not take in the impressions necessary to work a change in his motives and in his nature. . . . It tried to substitute religious teaching for the slow processes of nature, in the development of a people from barbarism to civilization. As well might they have attempted to make the Indian white by the use of soap and water (Scott, 1924, p. 116).

Missionary work in the Spokane Valley followed a similar course. In 1819 a Catholic French Canadian couple staying at the Spokane House became the parents of the first white child born in the Spokane Valley. With a growing interest to learn more about the white man's religion, the tribe selected two Spokane boys,

Spokane Gary and Pelly, to attend the Catholic missionary school, St. Boniface located in Red River, Manitoba, Canada, with the white child in 1825. Upon his return from St. Boniface in 1830, Spokane Gary taught his people what he learned about the Christian religion, holding simple services of prayers and hymns regularly (Jessett, 1960). Pelly, however, returned in poor health and died in 1831. Spokane Gary continued his work and also served as an interpreter and guide for the Protestant and Catholic missionaries that followed (Jessett, 1960; Wynecoop, 1969).

In 1838, Eells, Walker, and their families established a Protestant mission in Tshimakin, near present-day Ford, Washington. Spokane Gary adjusted his teaching of Christianity to fit the tribe's lifestyle, but this approach was frequently at odds with the missionaries. At the same time, the Spokanes became more interested in learning about the white man's religion from the white men. Even though none of the Spokanes converted to the white men's religion, Spokane Gary's influence diminished. At the same time, tensions between the increasing numbers of white settlers and the tribes continued to rise (Jessett, 1960).

In 1847, many members of the Cayuse Tribe in the southern Spokane Valley died from measles during an epidemic that swept through the region. The Cayuse believed the Whitmans, Presbyterian missionaries from New York, introduced the deadly disease to their tribe. In response, Cayuse warriors attacked the Whitman Mission in Walla Walla, killing the Whitmans along with 13 or more Americans (Johnson, 1904). Out of concern for their safety, Eells and Walker decided to abandon the Tshimakain Mission in May of 1848. Spokane Gary once again regained his influence as a leader among his people. It was not until 1866 when Chief Baptiste Peone of the Upper Spokanes requested the services of the Catholic priest Father Joseph Cataldo, that others attempted to establish missions among the Spokanes (Wynecoop, 1969).

Early Laws and the Indian Removal Act of 1830

As colonization spread, an increasing numbers of white traders and squatters encroached upon important hunting and gathering grounds in the Willamette and Spokane valleys. The physical and cultural survival of its indigenous inhabitants became more challenging. These circumstances imposed the development of new relationships between the tribes and the American government. While most tribes did not have political loyalties beyond their own band, the Americans treated the tribes as miniature sovereign nations residing within United States boundaries (White, 1991). At the same time, the United States Constitution justified the Americans control over tribal affairs. Through this document, 'Congress has plenary authority to legislate for the Indian tribes in all matters, including their form of government' (Pevar, 1992, p. 48). Congress interpreted the Constitution to include dictating compensation for desirable lands through treaties and forcing tribes to migrate from aboriginal lands. To that end, Congress created a number of laws and legislative acts to orchestrate and manage land acquisitions. The Land Ordinance of 1785 specified a system for land allocation to settlers such that:

After Indians ceded title to their lands to the federal government, surveyors would mark the land off into giant squares, six miles wide on a side and then subdivide them into sections of one square mile. Each Section, in turn, would contain quarter sections of 160 acres each. The government would then sell this land, a tier of townships at a time, at public auction. Any land left unsold after the auction could be purchased at the land office at a minimum price originally set at $2.00 an acre but reduced to $1.25 an acre in 1820 (White, 1991, p. 137-8).

To implement this system, the Ordinance of 7 August 1786 divided Native American lands into two geographical units, north and south. Each unit had a superintendent and two deputies authorized to negotiate land acquisitions between indigenous people and settlers in the east. In 1787, Congress extended application of this ordinance to other states and territories, including the Pacific Northwest (Harmon, 1941; Prucha, 1962).

In addition, the first Congress passed a number of laws in 1790 'to protect Indians from non-Indians,' such as licensing trade and prohibiting non-Indians from taking Indian land without consent (Pevar, 1992, p. 3). In some ways, the loose organization between tribes made them dependent on the United States government to protect their rights. As White (1991) suggests, the Untied States acted as if the Native Americans were minors under their guardianship. However, the United States government rarely enforced the laws.

Under Andrew Jackson's administration, federal Native American policy changed suddenly and dramatically. For example, the 1830 Indian Removal Act authorized the President to negotiate the relocation of eastern tribes to the West (Pevar, 1992). Not only did this increase the conflict between white settlers and indigenous tribes, but it also placed the eastern and western tribes at odds with each other. In efforts to manage these contentious affairs, Congress created the Office of Indian Affairs in 1833, reorganizing it into the Bureau of Indian Affairs in 1834 (Harmon, 1941). Little did the tribes of the West know that this was only the beginning of a new land 'sharing' era.

To make matters worse, the whites brought a host of new diseases to the Willamette Valley including measles, influenza, small pox, syphilis, and whooping cough. A small pox epidemic between 1792 and 1783 fatally infected up to half of the indigenous populations it struck and left its survivors severely scarred (Mackey, 1974). The 'intermittent fever' epidemic of 1830-1833, later thought to be either malaria imported from the Sacramento Valley of California or viral influenza, claimed the lives of five to six thousand Native Americans along the banks of the Willamette River (Cook, 1952; Taylor and Hoaglin, 1962). In all, about 9 out of 10 indigenous people in the Willamette Valley died from imported diseases before 1840. As Mackey (1974) crudely put it, 'disease helped make possible the settlement of the Willamette Valley by the whites, almost without resistance, and settlement completed the collapse of the Indian culture' (p. 21).

The Preemption Act of 1841

The first congressional act that specifically promoted colonization of the West was the Preemption Act. This act encouraged the establishment of small farms by allowing settlers to purchase up to 160 acres for $1.25 per acre after living on it for 14 months (Bohm and Holstine, 1983). The only settlers that could benefit from this act, however, were American citizens or persons intending to become American citizens. To qualify for American citizenship at this time, one had to be a free white person, of sound moral character, residing in America for at least 14 years (Immigration and Naturalization Service, 2002). In practical terms, the Preemption Act allowed qualified squatters to legally acquire title for lands they already occupied (White, 1991). Yet when Congress passed this act, no agreements about how to integrate the white settlers into the Oregon Territory existed between the United States and its indigenous inhabitants. Thus, as farmland replaced traditional hunting and gathering grounds, tensions between whites and the tribes increased.

In 1842, a group of 109 men, women and children arrived in the Willamette Valley. This marked the first of many large, organized emigrations. Approximately eight hundred newcomers came in 1843. By 1845, about 5,000 Americans resided in the Willamette Valley (White, 1991). Squatter occupation increased tensions not only between the tribes and the Americans, but also between the Americans and the British. As a result, in 1846 Britain signed a treaty relinquishing their land claims below the 49th parallel to the Americans but retained their claim to present-day Canada. This treaty formalized the American claim and established the boundaries to the Oregon Territory, which included the present-day states of Oregon, Washington, Idaho, western Montana and western Wyoming (Johnson, 1904; Clark, 1927). Still, no agreements about white settler occupation on indigenous lands were in place between the Americans and the tribes. This fact however, did not deter white settlement.

As the Oregon white population grew, so did an interest in provisional government and jurisdictional land subdivisions. On 15 June 1846, Congress granted the Oregon Territory state sovereignty (Johnson, 1904). Shortly thereafter, the subdivision of the Willamette Valley into counties began. On 28 December 1847, the state approved the boundaries of Linn County, named after Louis F. Linn, an advocate for what became known as the Donation Act (Clark, 1927; Linn County Pioneer Memorial Association, 1979). The following year, brothers Thomas and Walter Monteith founded the City of Albany, named after their hometown, Albany, New York (Victor, 1872). In 1851, the provisional government formally designated Albany as the county seat of Linn County (Mullen, 1971). Prior to its independent incorporation in 1974, Millersburg was formerly the area of Albany zoned for heavy industrial development. By 1849, the white population in the Willamette Valley reached 8,779 with 923 whites in Linn County alone (Bowen, 1978).

Later, subdivisions within the Oregon Territory created the territories of Washington (1853), Idaho (1863), Montana (1864) and Wyoming (1868) (White,

1991). In 1858, Washington's territorial government approved the administrative boundaries for Stevens County, home to Spokane Indian Reservation (designated in 1881) and Ford (founded in 1912). Stevens County, named after the first governor of the Washington territory, Isaac Stevens, originally encompassed Adams, Chelan, Douglas, Ferry, Franklin, Lincoln, Okanogan, Pend Oreille, Spokane, Stevens and Whitman counties. The final restructuring of Stevens County to the administrative boundaries of today, however, did not take place until 1911 (Stevens County Rural Development Planning Council, 1961). With established boundaries came more white settlers who negatively impacted and limited access to the Spokanes' traditional hunting, fishing and food gathering domains.

The Donation Act of 1850

The Donation Act was the second major congressional act to spur colonization of the West. This act donated public lands, full title and free of charge, to white citizens that were willing to settle the new west. Under this act, white citizens (including Americans with one indigenous parent and eligible persons declaring their intention to become American citizens), age 18 or older, residing in the Oregon Territory by 1 December 1850 (later modified to before 1 January 1855), could claim up to 320 acres of vacant, non-mineral land. Qualified married couples could claim 640 acres (Linn County Pioneer Memorial Association, 1979; Mullen, 1971). In both instances, this was twice the amount of land citizens could claim in other regions (Jessett, 1960). Southern states raised many concerns about how such generous allotments would detrimentally reduce the workforce that was necessary for their agricultural industry (Prucha, 1962). In response, on 1 December 1853 Congress limited claims to 160 acres for single persons and 320 acres for married couples (Linn County Pioneer Memorial Association, 1979). In order for a Native American to keep the land he or she already occupied, he or she had to sever tribal affiliations and become an American citizen. Because they did not understand or were not aware of the law, many Native Americans lost the lands they had inhabited for generations along with the livestock they maintained on it (Wynecoop, 1969).

By 1850, development of the Willamette Valley was well underway with the white population reaching 11,873 (Bowen, 1978). Albany's first post office and its first school opened their doors in 1850 and 1851, respectively. In 1859, G.H. Hackleman published Albany's first newspaper, the *Oregon Democrat*, now called the *Democrat-Herald*. The town and its industrial center continued to grow with little resistance from the dwindling Willamette Valley tribes. By 1860, the Willamette Valley white population soared to over 30,000. There were 6,763 whites in Linn County alone (Mullen, 1971).

Acquisition of the Spokane Valley lands, however, came with resistance. Isaac Stevens, first governor of the Washington territory, was more eager to acquire land holdings than he was to secure real Native American consent. His rushed tactics cultivated discontent among the tribes (White, 1991). Stevens met with the

Spokanes for the first time in December of 1855 to establish a treaty but they did not reach an agreement (Wynecoop, 1969).

As word of the first gold strike near Colville in 1854 spread, white settlers flocked to the upper Spokane Valley. In addition to white settlers, Chinese miners frequently moved onto abandoned placer claims and reworked the tailings. The white population of the state of Washington was a mere 1,201 in 1850. By 1860, Washington's population jumped to 11,594 whites, with 996 whites in the Spokane Valley alone. As whites continued to settle in the area despite the absence of a treaty, indigenous food supplies continuously declined and significantly threatened the Spokanes' way of life. Injurious conflicts between the Native Americans and whites ensued.

In 1858, the Spokanes heard that Colonel Steptoe and 150 soliders were coming to Colville to investigate the murder of two miners. In response, the Spokanes made alliances with the Coeur d' Alenes, Yakimas, Kalispels and Palouse to defend their aboriginal lands. The alliance forced Steptoe and his troops to retreat near present-day Rosalia. Colonel Wright then led a retaliatory campaign of 700 well-armed soldiers and defeated the Spokanes. Following his victory, Wright rounded up the remaining horses of the Spokane and slaughtered them. He continued his rampage, destroying all standing crops along the way and hanging 15 Spokane for alleged murders at a site now known as Hangman's Creek. Having lost most of their food stores, many Spokane starved to death the following winter. Efforts to establish a treaty and place the remaining Spokanes onto a reservation resumed but to no avail (Wynecoop, 1969).

Grand Ronde Reservation

In an attempt to halt conflicts while simultaneously opening lands to white settlement, reservations became a means for segregating Native Americans and whites. Frequently lands allotted for reservations were undesirable grounds lying outside of traditional hunting and gathering areas. At best, reservation lands were mere fragments of previous indigenous domains. With the loss of land, came the loss of cultural traditions. This rendered many tribes virtually powerless. Reservations were 'above all, supposed to be a place where Indians were to be individualized and detribalized . . . the means to destroy the tribes as political entities' (White, 1991, p. 92). Ironically, treaty negotiations legally recognized tribes as sovereign nations, dictating government-to-government relationships. While support for the reservation system prevailed, not all Willamette Valley residents backed its policies. Editorials submitted to *The Oregonian* newspaper in 1878 and 1880, respectively, depict such disagreements:

> Our Indian policy is irrational, and very grave errors have been committed in carrying it into effect. We have established numerous reservations, yet do not fully support the Indians assigned to them; while the advancing tide of immigration fills the country round about them, destroys the natural food supply and prevent the Indian from supporting himself (Scott, 1924, p. 133).

Individual man is the prime factor. We shall never make any progress worth notice with the Indians so long as we essay to keep them in a common herd; for the primary condition of all human advancement is the effort and the growth of the individual man (Scott, 1924, p. 135).

These sentiments did not however, deter white settlement or development.

After considerable depletion by disease, the surviving tribes of the Willamette Valley signed a treaty in 1855. In exchange for their indigenous lands, the tribes received $62,260 and the Grand Ronde Indian Reservation. The reservation encompassed approximately 56,699 acres northwest of Albany, Oregon, and about 25 miles from the Pacific Ocean in Polk County. In 1858, one year after Congress ratified the treaty, a melting pot of about 1,200 Native Americans from more than 20 different tribes and bands lived on the reservation (Clark, 1927). Reservation life began with leaving all of one's personal possessions behind and starting over with nothing. The population density on the reservation was greater than any of its residents had ever known given the following land allocation scheme:

. . . to a single person over twenty one years of age, twenty acres; to a family of two persons, forty acres; to a family of three and not exceeding five persons, fifty acres; to a family of six persons and not exceeding ten, eighty acres; and to each family over ten in number, twenty acres for each additional three members (Mackey, 1974, p. 139).

Residents could leave the reservation only under work permit stipulations. Schools required children to cut their hair short, wear uniforms and refrain from practicing tribal customs (Mackey, 1974). Disguised as protecting the Native Americans and preparing them for integration into white society, these activities caused tremendous cultural alienation. According to the 1870 report of Charles LaFollett, the Indian Agent managing the Grand Ronde Reservation, the end result was successful assimilation of the Willamette Valley tribes:

No man can visit this agency and go away without being impressed with the wonderful improvement of these Indians. They are marching along, not slowly, but with rapid strides to civilization. Less crime has been committed by them in the past year than by the same number of whites. Not a drunken Indian has been seen on the agency in a year. Not an Indian has been whipped since I have been in charge, and but one in the guard-house, and he only for two days. Yet the discipline is all that could be desired (Bureau of Indian Affairs, 1870, p. 56).

Homestead Act of 1862

The third significant congressional act to encourage colonization of the West was the Homestead Act of 1862. Under the Homestead Act, American citizens and persons intending to become citizens could claim 160 acres. After improving and residing on the land for five years, those eligible could gain full title to the land for a registration fee of $26.00 (Bohm and Holstine, 1983). The Homestead Act was perhaps the single most successful colonization plan. Citizens claimed roughly

270 million acres, or ten percent of the United States, through this program (National Parks Service, 2002). The Spokane indigenous lands were difficult to clear and cultivate so many farmers rented their land out in order to make ends meet (Bohm and Holstine, 1983). Perhaps the difficult terrain of the upper Spokane Valley spared the area from the rapid development seen in the Willamette Valley. By 1870, the white population of the Spokane Valley dropped to 734 whereas the Willamette Valley jumped to 44,488 with Linn County reaching 8,717, nearly 2,000 of whom lived in the city of Albany (Bohm and Holstine, 1983; Bowen 1974; Clark, 1927).

Spokane Reservation

While colonization of the Willamette Valley was well underway in the 1870's, formal agreements opening the Spokane Valley to white settlers were absent. Over the next two decades, however, that too, would change. In 1880, The United States Army built Fort Spokane to protect white settlers from possible Native American attacks. At that time, the Spokanes were roughly 3,000 in number (Wynecoop, 1969). In efforts to minimize tensions and gain control over the Spokanes, President R.B. Hayes signed an Executive Order on 18 January 1881 that established the Spokane Indian Reservation. The boundaries of the reservation encompassed 154,602.57 acres of land not considered the most desirable for food gathering, hunting and fishing. Due to lack of payment for aboriginal lands and religious differences among the upper, middle and lower Spokane bands, the upper and middle bands refused to settle on the reservation (Ruby and Brown, 1970).

A gold discovery near Coeur d' Alene in 1883 however, fueled a second significant gold rush. This increased the numbers of white settlers encroaching on Spokane aboriginal lands. In fear of losing their remaining land, the Spokanes entered an agreement on 18 March 1887 to move to the Spokane Reservation or other nearby reservations. In exchange for the move, the tribe received $127,000, earmarked for the erection of houses, purchase of cattle, seeds and farm equipment. Congress ratified the agreement on 13 July 1892. By 1897, 340 Lower Spokanes, and 188 Upper and Middle Spokanes lived on the Spokane Reservation (Wynecoop, 1969). Establishment of the reservation did not, however, offer full protection of this small slice of aboriginal land.

Railroad

The first train whistles of the Oregon Central Railroad blew in Albany, Oregon in 1871 (Victor, 1872). With fifteen water-powered manufacturing plants in operation by 1875, Albany established itself as the industrial hub of the Willamette Valley. The city's foundries, black smith shops, tanneries, creameries, flour and flax mills, furniture factories, carriage factories and sawmills processed the area's crops and timber, and supplied its inhabitants with amenities comparable to those of Portland (Mullen, 1971).

In May of 1889, the Spokane Falls and Northern Railroad Company started laying track, connecting Spokane to Colville. This initiated another substantial mining stampede that lasted through the 1890's. Hopeful prospectors filed nearly 12,500 mining claims by 1900. The timber industry also witnessed tremendous growth, adding more than 20 saw mills along the rail line between 1890 and 1892. By 1900, there were 102,696 persons residing in Spokane Valley. Among them, 10,543 lived in Stevens County (Stevens County Rural Development Planning Council, 1961; Wynecoop, 1969).

General Allotment Act of 1887

In efforts to complete colonization of the West, Congress passed the General Allotment Act (i.e., the Dawes Act) in 1887. Its primary purpose was 'to break up tribal government, abolish Indian reservations, and force Indians to assimilate into white society' (Pevar, 1992, p. 5). Through this act, surveyors divided reservation lands into allotments ranging from 80 to 160 acres, just a fraction of what the first white settlers in the Oregon Territory could claim under the Donation Act. Each tribal member age 18 or older could claim one allotment. After all eligible tribal members filed claims, the United States government designated the remaining allotments as surplus and sold them to interested white settlers (Tyler, 1973). This reduced the 140 million acres collectively owned by Native Americans in 1887 to a mere 50 million acres by 1934 when Congress abolished the act (Pevar, 1992). This act significantly impacted both the Grand Ronde and Spokane Reservations.

Following the completion of a land survey on the Grand Ronde Reservation in 1901, members of the Willamette Valley tribes claimed 270 allotments covering approximately 31,146 acres. White settlers then purchased the 25,791 acres declared surplus (Confederated Tribes of the Grand Ronde Community, 2002). By 1920, over 200,000 lived in the Willamette Valley. The Linn County population grew to 24,550. The following year, the last Calapooian died (Stanard, 1948), just three years before all Native Americans became citizens of the United States.

The Spokane Tribe's control over their reservation lands started to change prior to the completion of a land survey. The Act of 3 March 1905 authorized the Secretary of the Interior to sell reservation lands along the Spokane River for the development of hydroelectric power. Through this program, David Wilson constructed the Little Falls Power Plant in 1906. That same year, allotment of lands on the Spokane Reservation began. Allotment sizes ranged from 80 acres for farm tracks, 120 acres for mixed agriculture and grazing, and 160 acres for grazing and timber (Wynecoop, 1969). In all, tribal members claimed 651 allotments, covering roughly 65,000 acres of the 154,602.57-acre reservation. Impatient white settlers began squatting on the unclaimed parcels the following year even though land declared as surplus was not made available for purchase until 1909. In spite of accusations that 'the Indians gobbled up the best land,' 40,000 individuals filed claim applications within the first 8 days of opening purchasing (Wynecoop, 1969, p. 36). Under the Railroad Land Grant of 1866, the Northern Pacific Railroad claimed every odd section of land along the track that ran through the reservation. Their claim totaled 55,000 acres but was reversed 5 years later (Ruby and Brown, 1970).

By 1910, Stevens County housed 100 saw mills and grew to a total county population of 25,297 (Bohm and Holstine, 1983). Feeding the growth frenzy, on 9 April 1910, Congress opened the Spokane Reservation for mineral leasing and extraction. White prospectors, however, failed to identify ore fortunes on the Spokane Reservation. Instead, two Spokane tribal members made the first sizable lode discovery, the Midnite Mine, nearly 45 years later (Wynecoop, 1969). Another big blow to the Spokanes' culture came with the Act of 29 June 1940. This act authorized the acquisition of reservation land along the Spokane River for the construction of the Grand Coulee Dam and Reservoir. The 550-foot dam put an end to both historical salmon runs and all of the tribe's traditional ceremonies associated with them.

The Termination Act of 1954 and Other Acts

The final step to fully integrate Native Americans into white society was the Termination Act of 1954. This act authorized Congress to sever trust relationships with tribes. Practically speaking, this meant terminating all federal benefits and services offered to tribes and eliminating tribal land holdings (Pevar, 1992). Congress terminated 109 tribes, 64 of who were indigenous to Oregon, between 1954 and 1966 (Oregon Legislative Commission on Indian Services, 1999). While the Spokane Tribe successfully resisted termination, the tribes of the Willamette Valley did not. Congress restored the federal benefits and services to several tribes between 1973 and 1977. However, it was not until 1983 with the passing of the Grand Ronde Restoration Act that Willamette Valley tribal ancestors regrouped under the Confederated Tribes of the Grand Ronde Community. When President Ronald Reagan signed the Grand Ronde Reservation Act in 1988, the Confederated Tribes of the Grand Ronde regained 9,811 acres of the original Grand Ronde Reservation. Since then, this young confederation has focused its efforts on rebuilding tribal institutions and services for more than 4,700 tribal members. Striving towards self-sufficiency, the Confederated Tribes of the Grand Ronde Community opened the Mountain Spirit Casino in 1995 (Confederated Tribes of the Grand Ronde Community, 2002).

The political power of the Spokanes began to improve in the 1950's. Under the Indian Reorganization Act of 1934, the Spokane Tribe developed and approved its own constitution and bylaws on 27 June 1951. Less than two months later, the tribe filed a claim against the United States government for underpayment of their lands. A second docket alleged that the United States government mismanaged monies and properties that it held in trust for the tribe. The final judgement for these two claims rendered on 21 February 1967 awarded $6.7 million to the tribe (Wynecoop, 1969). In 1981, additional claims alleging mismanagement of tribal assets resulted in an award of $271,431.23. The enactment of Public Law S5-240 in 1958, returned 2,752.35 acres of vacant and indisposed land declared surplus under the Donation Act to the Spokanes. While forever changed by the impacts of colonization, the tribe's 2,153 members work together to preserve their culture and to achieve self-sufficiency (Spokane Tribe of Indians, 1999).

Summary

Colonization removed all of the indigenous peoples from the Willamette Valley. Moreover, it removed a Native American voice from all land use decisions in Linn County and its towns and cities. The Spokane Tribe continues to occupy a small section of their original hunting and gathering domains. Increased population density and land use restrictions, however, prohibit complete adherence to traditional customs and lifestyles. This creates an atmosphere where maintenance of culture competes with the need for survival and requires the development of new relationships with the land. Decision-makers must be aware of these challenges and how they influence current land use beliefs. Failure to do so may make decision-making processes contentious.

Appendix C

Hazards Associated with
Teledyne Wah Chang Albany Operations

Table C.1 Teledyne Wah Chang Albany (TWCA) Hazard Ranking
     ·   **System Scores**

HRS Pathway Scores	10 August 1982	15 September 1982	22 February 1983	2 June 1983	NPL Final 12 August 1983
Ground Water Route	35.92	35.92	44.90	44.90	44.90
Surface Water Route	21.82	58.18	26.67	40.97	47.27
Air Route	71.92	71.92	0	67.56	67.56
Overall HRS Score	48.15	57.36	30.19	52.53	54.27
Fire and Explosion Potential	0	0	0	0	0
Direction Contact Potential	33.33	33.33	33.33	33.33	33.33
Inspector	J. Betz	J. Betz	H. Aldis	J. Betz	J. Betz

Source: Environmental Protection Agency, Site Listing Docket for the Teledyne Wah Chang Albany Superfund Site, Seattle, WA.

Table C.2 Organic substances in groundwater at the TWCA Superfund Site

Location	Maximum Concentration of Organic Substances in TWCA Ground Water
Farm Ponds Area	VOLATILE ORGANICS: 1,1-dichloroethane (120 ppb), 1,1-dichloroethylene (4J ppb), 1,2-dichloroethane (6 ppb), ,2-dichloroethylene (52 ppb), 11,1,2,2-tetrachlorethane (5 ppb), tetrachloroethylene (130 ppb), 1,1,1-trichloroethane (13 ppb), 1,1,2-trichloroethane (41 ppb), trichloroethylene (66 ppb), vinyl chloride (11 ppb).
Extraction Area	VOLATILE ORGANICS: acetone (230 ppb), benzene (62 ppb), carbon disulfide (92J ppb), chloroform (52 ppb), 1,1-dichloroethane (140 ppb), 1,1-dichloroethylene (110 ppb), 1,2-dichloroethylene (36 ppb), methylisobutylketone (7,500 ppb), tetrachloroethylene (19J ppb), 1,1,1-trichloroethane (600 ppb), trichloroethylene (330 ppb).
Fabrication Area	VOLATILE ORGANICS: acetone (3,400 ppb), benzene (60 ppb), chloroethane (420D ppb), chloroform (27J ppb), 1,1-dichloroethane (4,200 ppb), 1,1-dichloroethylene (24,000 ppb), 1,2-dichloroethane (220J ppb), methylisobutylketone (85,000D ppb), tetrachloroethylene (150 ppb), 1,1,1-trichloroethane (45,000 ppb), 1,1,2-trichloroethane (5 ppb), trichloroethylene (910J ppb), vinyl chloride (50 ppb), xylenes (46 ppb).
Solids Area	VOLATILE ORGANICS: acetone (21 ppb), 2-butanone (12 ppb) chloroform (5 ppb), 1,1-dichloroethane (23 ppb), 1,1-dichloroethylene (8 ppb), 1,2-dichloroethane (43 ppb), methyl chloride (4J ppb), methylisobutylketone (7J ppb), tetrachloroethylene (2J ppb), 1,1,1-trichloroethane (22 ppb), 1,1,2-trichloroethane (1J ppb), trichloroethylene (29 ppb), vinyl chloride (11 ppb).
	SEMIVOLATILE ORGANICS: bis (2-ethylhexyl)phthalate (130 ppb), di-n-butylphthalate (2J ppb), di-n-octyl-phthalate (10 ppb).

J = Estimated value below method detection limits. ppb = parts per billion.

Source: Environmental Protection Agency, 1989, 1994b, 1995.

Table C.3 Inorganic substances in groundwater at the TWCA Superfund Site

Location	Maximum Concentration of Metals and Radionuclides in TWCA Groundwater
Farm Ponds Area	aluminum (154,000 ppb), barium (958 ppb), cadmium (25.8 ppb), calcium (1,020,000 ppb), chromium (240 ppb), copper (110 ppc), iron (139,000L ppb), lead (41 ppb), magnesium (479,000 ppb), manganese (3,460 ppb), nickel (152 ppb), sodium (368,000 ppb), thallium (5.5 ppb), thorium (30.4L ppb), zinc (6,270 ppb), radium-226 (2.2 pCi/l), radium-228 (3.4) pCi/l).
Extraction Area	aluminum (378,000 ppb), arsenic (234 ppb), barium (1,080 ppb), cadmium (9.2 ppb), calcium (420,100 ppb), chromium (373 ppb), copper (289 ppc), iron (472,000 ppb), lead (62.8L ppb), magnesium (280,000 ppb), manganese (20,900 ppb), nickel (171 ppb), potassium (20,220 ppb), sodium (500,000 ppb), thorium (30 ppb), zinc (638 ppb), uranium (53.8 ppb), radium-226 (2.2 pCi/l), radium-228 (3.4) pCi/l).
Fabrication Area	aluminum (990,000 ppb), arsenic (107L ppb), barium (3,310 ppb), cadmium (31.6 ppb), calcium (426,000 ppb), chromium (614 ppb), copper (3,920 ppc), iron (630,000 ppb), lead (180 ppb), magnesium (235,500 ppb), manganese (34,000 ppb), nickel (2,620 ppb), selenium (53 ppb), silver (40K ppb), sodium (500,000 ppb), thallium (5.1 ppb), thorium (183L ppb), tin (168B ppb,)zinc (1,230 ppb), uranium (250 ppb), radium-226 (8.4 pCi/l), radium-228 (31 pCi/l).
Solids Area	aluminum (282,000 ppb), antimony (18.1B), arsenic (21.8L ppb), barium (2,800 ppb), cadmium (40.7 ppb), calcium (1,990,000 ppb), chromium (405 ppb), copper (269 ppc), iron (504,000 ppb), lead (61.6J ppb), magnesium (11,400,000 ppb), manganese (72,500 ppb), nickel (660 ppb), selenium (9.5K ppb), sodium (1,160,000 ppb), thallium (5.5L ppb), thorium (45.7 ppb), tin (2,980 ppb), zinc (670L ppb), uranium (16.2 ppb), radium-226 (8.5 pCi/l), radium-228 (4.2 pCi/l).

J = Estimated value below method detection limits. K = Biased high. L= Biased Low.
ppb = parts per billion. pCi/l = picocuries per liter.

Source: Environmental Protection Agency, 1989, 1994b, 1995.

Table C.4 Substances in surface water at the TWCA Superfund Site

Maximum Concentration of Surface Water Contaminants
VOLATILE ORGANICS: acetone (190 ppb), 1,1-dichloroethane (6J ppb), 1,1-dichloroethylene (3J ppb),   1,2-dichloroethylene (3J ppb), methylisobutylketone (990 ppb),   1,1,2,2-tetrachlorethane (1J ppb), toulene (29 ppb),   1,1,1-trichloroethane (6 ppb), trichloroethene (5J ppb).
SEMIVOLATILE ORGANICS: bis (2-ethylhexyl)phthalate (6J ppb), dietylphthalate (6J ppb).
METALS AND RADIONUCLIDES: aluminum (35,200 ppb), arsenic (15.3 ppb), barium (479 ppb),   cadmium (0.6L ppb), calcium (1,320,000 ppb), chromium (84 ppb),   copper (34 ppb), lead (45 ppb),magnesium (84,800 ppb), nickel (55.2 ppb),   potassium (14,800 ppb), selenium (8.6 ppb), silver (1.4L ppb), sodium   (301,000K ppb), thorium (3.4 ppb), zinc (22 ppb), uranium (19.4 ppb),   radium-226 (1.5 pCi/l), radium-228 (2.7) pCi/l).

J = Estimated value below method detection limits. ppb = parts per billion.
pCi/l = picocuries per liter.

Source: Environmental Protection Agency, 1989, 1994b, 1995.

Table C.5 Substances in main plant containment ponds at the TWCA Superfund Site

Location	Maximum Concentration of Contaminants in Solids
Schmidt Lake Solid Materials	VOLATILE ORGANICS: methylene chloride (0.09 mg/kg), 1,1,1-trichloroethane (0.32 mg/kg), 4-methyl-2-pentanone (54 mg/kg), 1,1-dichloroethane (3.9 mg/kg), tetrachloroethene (0.073 mg/kg).
	SEMIVOLATILE ORGANICS: hexachlorobenzene (25.33 mg/kg), bis(2-ethyl-hexyl)phthalate (1 mg/kg), n-nitroso-di-n-propylamine (0.59 mg/kg).
	METALS, RADIONUCLIDES AND OTHER INORGANICS: antimony (14 mg/kg), arsenic (36 mg/kg), barium (72 mg/kg), beryllium (1.1 mg/kg), cadmium (1.2 mg/kg), chromium (13 mg/kg), copper (72 mg/kg), cyanide (110 mg/kg), lead (150 mg/kg), mercury (1.4 mg/kg), nickel (4,300 mg/kg), selenium (4 mg/kg), zinc (97 mg/kg), zirconium (28.8 mg/kg), thorium (7.5 pCi/g), uranium (160.9 pCi/g), radium-226 (26.4 pCi/g).
Lower River Solids Pond Materials	VOLATILE ORGANICS: methylene chloride (22.0 mg/kg), 1,1,1-trichloroethane (0.86 mg/kg), 4-methyl-2-pentanone (1,400 mg/kg), 1,1-dichloroethane (0.86 mg/kg), tetrachloroethene (0.97 mg/kg).
	SEMIVOLATILE ORGANICS: hexachlorobenzene (64 mg/kg), bis(2-ethyl-hexyl)phthalate (1.7 mg/kg).
	METALS, RADIONUCLIDES AND OTHER INORGANICS: antimony (24 mg/kg), arsenic (39 mg/kg), barium (3,500 mg/kg), beryllium (1.3 mg/kg),chromium (220 mg/kg), copper (77 mg/kg), cyanide (165 mg/kg), lead (260 mg/kg), mercury (7.6 mg/kg) nickel (3,000 mg/kg), selenium (16 mg/kg), zinc (87 mg/kg), zirconium (14 mg/kg), thorium (8.3 pCi/g), uranium (87.8 pCi/g), radium-226 (22.2 pCi/g).

mg/kg = mlligrams per kilograms, dry weight.   pCi/g = picocuries per gram.

Source: Environmental Protection Agency, 1989, 1994b, 1995.

Table C.6 Substances in surface soils at the TWCA Superfund Site

Location	Maximum Concentration of Surface Soil Contaminants
Farm Ponds Area	SEMIVOLATILE ORGANICS AND PCS: hexachlorobenzene (2,000 ppb), Total PCBs (1.4 ppm) METALS AND RADIONUCLIDES: chromium (69 ppm), thorium (25 ppm), zirconium (13,500 ppm), radium-226 (8 pCi/g), radium-228 (3.8 pCi/g).
Extraction Area	SEMIVOLATILE ORGANICS: benzo(a)anthracene (870 ppb), benzo(a)pyrene (610 ppb), benzo(b)fluranthene (870 ppb), benzo(k)fluranthene (1,100 ppb), chrysene (1,200 ppb), dibenzo(a, h)anthracene (140 ppb) hexachlorobenzene (8,000 ppb), indeno(1,2,3-cd)pyrene (400 ppb), total PCBs (19 ppm).
	METALS AND RADIONUCLIDES: chromium (1,010 ppb), thorium (69.9 ppb), zirconium (198,000 ppm), radium-226 (17.9 pCi/g), radium-228 (5.9 pCi/g).
Fabrication Area	SEMIVOLATILE ORGANICS: benzo(a)anthracene (1,700 ppb), benzo(a)pyrene (1,300 ppb), benzo(b)fluranthene (1,400 ppb), benzo(k)fluranthene (1,500 ppb), chrysene (2,000 ppb), dibenzo(a, h)anthracene (250 ppb), hexachlorobenzene (5,100 ppb), indeno(1,2,3-cd)pyrene (970 ppb), total PCBs (9.2 ppm).
	METALS AND RADIONUCLIDES: chromium (2,810 ppb), thorium (13.1 ppb), radium-226 (5 pCi/g), radium-228 (11.6 pCi/g).

ppb = parts per billion. ppm = parts per million. pCi/g = picocuries per gram.

Source: Environmental Protection Agency, 1989, 1994b, 1995.

Table C.7 Substances in subsurface soils at the TWCA Superfund Site

Location	Maximum Concentration of Subsurface Soil Contaminants
Farm Ponds Area	SEMIVOLATILE ORGANICS AND PCBs: hexachlorobenzene (2.4 ppb), total PCBs (.04 ppm).
	METALS AND RADIONUCLIDES: thorium (13.6 ppm), radium-226 (1.7 pCi/g), radium-228 (1.6 pCi/g).
Extraction Area	SEMIVOLATILE ORGANICS: hexachlorobenzene (670 ppb), total PCBs (13 ppm).
	METALS AND RADIONUCLIDES: thorium (75 ppb), radium-226 (54.2 pCi/g), radium-228 (11.43 pCi/g).
Fabrication Area	SEMIVOLATILE ORGANICS: benzo(a)anthracene (1,100 ppb), chryscne (1,300 ppb), hexachlorobenzene (27,000 ppb), total PCBs (440 ppm).
	METALS AND RADIONUCLIDES: thorium (170 ppm), radium-226 (26 pCi/g), radium-228 (6.2 pCi/g).

ppb = parts per billion. ppm = parts per million. pCi/g = picocuries per gram.

Source: Environmental Protection Agency, 1989, 1994b, 1995.

Table C.8 Half-life and Emitter Types of Radioactive Elements Present

Element	Half-life	Type of Emitter
Uranium-234	240,000 years	Alpha, Beta
Uranium-235	700 million years	Alpha, Gamma
Uranium-238	4.5 billion years	Alpha
Thorium-230	77,000 years	Alpha
Thorium-232	14.1 billion years	Alpha
Radium-226	1,600 years	Alpha, Gamma
Radium-228	5.8 years	Beta
Lead-210	22 years	Beta, Gamma

Source: Department of Energy 1989, American National Libraries 2001.

Appendix D

Hazards Associated with
Dawn Mining Company Operations

Table D.1 Annual Precipitation at the Midnite Mine and Surrounding Area

Data Source	Annual Precipitation
DeGuire 1985	20 inches annually (p. 17); 1931-1960 = 20.17 inches; 1961-71 = 17.03 inches (p. 18).
Shepherd Miller, Inc 1991	1949-1989 = average of 20 inches annually (9).
Sumioka 1991	19.4 inches per year at Wellpinit from 1951-1980 (p. 7).
Marcy et al. 1994	18.6 inches (473 mm) per year at Wellpinit from 1951-80 (p. 2).
US Bureau of Mines 1994	19 inches (48 cm) per year for 1951-80 (p. 3).
USGS 1996	19.4 per year for 1951-1980 (p. 2).
Superfund Technical Assessment and Response Team 1998	At mine site: 17.3 inches per year since 1993; at Wellpinit: 19.4 inches per year (p. 6-2).
URS Corporation 2000	17.5" annually (p. 2-2).

mm = millimeters. cm = centimeter.

Table D.2 Distance between Midnite Mine, DMC Mill site, and Spokane, Washington

Data Source	Distance from Spokane
EPA NPDES Permit 1986	Midnite Mine is 35 miles northwest Spokane (p. 2).
Shepherd Miller, Inc. 1991	Midnite Mine is 8 miles northwest of Wellpinit (p. 6).
Sumioka 1991	Midnite Mine is 40 miles northwest of Spokane (p. 3), 7 miles west of Wellpinit (p. 7).
Washington State Department of Health 1991	Midnite Mine is 18 miles west-northwest of the mill site (p. 1-1). The DMC mill site is 40 miles northwest of Spokane (p. 1-6).
Marcy 1993	Midnite Mine is 40 miles northwest, Spokane 8 miles northwest Wellpinit (p. 467).
Marcy et al. 1994	Midnite Mine is 40 miles northwest Spokane, 8 miles northwest Wellpinit (p. 2).
US Bureau of Mines 1994	Midnite Mine is 50 miles northwest Spokane (p. 2528).
BLM 1996	Midnite Mine is 45 miles northwest of Spokane, 25 miles west of Ford mill (p. 2528).
USGS 1996	Midnite Mine is 40 miles northwest of Spokane, 5 miles west of Wellpinit (p. 2).
Superfund Technical Assessment and Response Team 1998	Midnite Mine is 8 miles northwest of Wellpinit (p. 2-2).
URS Corporation 2000	Midnite Mine is 8 miles northwest of Wellpinit (p. 2-1).
EPA NPL Listing 2000d	Midnite Mine is 8 miles northwest of Wellpinit (p. 3.1-2).
URS Corporation 2001a	Midnite Mine is 35 miles from Spokane and 8 miles northwest of Wellpinit (p. 1).

Table D.3 Elevation of the Midnite Mine Site

Data Source	Elevation
DeGuire 1985	2,400 to 3,570 feet (p. 20).
Shepherd Miller, Inc 1991	2,400 to 3,570 feet (p. 6).
Marcy 1993	3,400 feet (1,036 m) to 2,400 feet (730 m) (p. 469).
Marcy et al. 1994	1,036 m (3,400 ft) to 730 m (2,400 ft) (p. 2).
US Bureau of Mines 1994	1,036 m (north end) to 730 m (south end) (p. 3).
BLM 1996	over 1,000 ft gain (p. 2528).
Superfund Technical Assessment and Response	3,171to 2,600 feet (p. 2-7); 2,400 to 3,400 (p. 6-1).
URS Corporation 2000	2,400 to 3,400 ft (p. 2-1).

m = meter

Table D.4 Description of the Midnite Mine Ore Body

Data Source	Ore Body
DeGuire 1985	Up to 600 feet wide, 1,200 feet long, 150 feet thick (p. 6).
Marcy 1993	Up to 60 feet (18 meters) wide, 1,200 feet (366 meters) long, 150 feet (46 meters) thick, 15 to 300 feet (4.6 to 91 meters) below the surface (p. 469).
Marcy et al. 1994	Up to 18 meters (60 feet) wide, 366 meters (1,200 feet) long, 46 meters (150 feet) thick, and 4.6 to 91 m (15 to 300 ft) below the surface (p. 7).
US Bureau of Mines 1994	Up to 18 meters wide, 366 meters long, 46 meters thick, and 4.6 to 91 meters below the surface (p. 4).
USGS 1996	Up to 55 feet wide, 1,100 feet long, 140 feet thick, and within in 285 feet of the present land surface (p. 5).
Superfund Technical Assessment and Response Team 1998	Up to 200 feet wide, 700 feet long, 150 feet thick, and 15 to 300 feet below ground (p. 2-7).
URS Corporation 2000	Up to 200 feet wide, 700 feet long, 150 feet thick, 15 to 300 feet below ground (p. 2-2).

ft = feet. m = meters.

Table D.5 Substances present in Pit 3 water at the Midnite Mine

Data Source	Pit 3 Water Characteristics and Contaminants
Shepherd Miller, Inc 1991	Effluent limits not exceeded except for manganese (p. 18); encountered a deep bedrock aquifer in Pit 3 (p. 14).
Sumioka 1991	Pit 3 pH = 4.5-4.7, exceeds EPA minimum (6.5) for fresh water aquatic life protection. Maximum concentrations: calcium (240-370 mg/l), sulfate (1,500-2,200 mg/l), magnesium (110-170), sodium (59-160 mg/l); selenium, beryllium, cadmium, copper, manganese, nickel, zinc, uranium, and radium-226 exceed EPA drinking water criteria (p. 30).
Marcy et al. 1994	Acidic water, containing several metals; water level is increasing in Pit 3, the deepest of the pits (p. 21).
Superfund Technical Assessment and Response Team 1998	Maximum concentrations: arsenic (16.3 µg/l), barium (48.2 JBK µg/l), beryllium (43.7µg/l), cadmium (47.1 µg/l), chromium (34.4 JK µg/l), cobalt (1,100 µg/l), copper (286 µg/l), lead (39.4 µg/l), manganese (101,000 µg/l), nickel (1,770 µg/l), selenium (65.3 µg/l), zinc (4,170 µg/l), uranium (21,800 µg/l), radium-226 (585 pCi/l), radium-228 (4.8 pCi/l), thorium-230 (170.8 pCi/l), thorium-232 (47.2 pCi/l), uranium-234 (6,610 pCi/l), uranium-235 (895 pCi/l), uranium-238 (5,980 pCi/l) (p. 5-3).
URS Corporation 2000	Pit 3 pH is 4.5-4.7 (p. 3-19); "well below the acid tolerance range for most fish species" (p. 3-20).

B: estimated, below contract required detection limits. J: estimated quantity.
K: unknown bias. mg/l = milligrams per liter. µg/l = micrograms per liter.
pCi/l = picocuries per liter.

Living in a Contaminated World

Table D.6 Substances present in Pit 3 sediments at the Midnite Mine

Data Source	Maximum Concentrations of Pit 3 Sediment Contaminants
Superfund Technical Assessment and Response Team 1998	arsenic (54.7mg/kg), barium (516 mg/kg), beryllium (7 mg/kg), chromium (19.6 mg/kg), cobalt (62.1 mg/kg), copper (102 mg/kg), lead (25.2 mg/kg), manganese (1,350 mg/kg), nickel (87.6 mg/kg), vanadium (26.6 mg/kg), zinc (46.1 mg/kg), uranium (917 mg/kg), radium-228 (4,900 pCi/kg), thorium-230 (139,200 pCi/kg), uranium-234 (284,900 pCi/kg), thorium-230 (139,200 pCi/kg), uranium-235 (7,900 JK pCi/kg), uranium-238 (305,000 pCi/kg) (p. 5-3).

J: estimated quantity. K: unknown bias.
mg/kg = mlligrams per kilograms. pCi/kg = picocuries per kilogram.

Table D.7 Substances present in Pit 4 water at the Midnite Mine

Data Source	Pit 4 Water Characteristics and Contaminants
Sumioka 1991	pH is 7.6 (p. 25), nitrate-nitrogen ratio exceeds EPA drinking water criteria (p. 24); beryllium, manganese, nickel, uranium and radium-226 exceed EPA drinking water criteria (p. 27).
Marcy et al. 1994	Similar composition to Pit 3 but lower metal concentrations and higher pH (p. 21).
Superfund Technical Assessment and Response Team 1998	Maximum concentrations : barium (7.8 JBK μg/l), chromium (2.3 JBK μg/l), cobalt (2.5 JBK μg/l), copper (23.1 JBK μg/l), lead (7.3 μg/l), manganese (840 μg/l), nickel (15.7 JBK μg/l), zinc (28.1 JK μg/l), uranium (2,640 μg/l), radium-226 (4.9 pCi/l), thorium-230 (12.9 pCi/l), thorium-232 (3.8 pCi/l), uranium-234 (824 pCi/l), uranium-235 (66 pCi/l), uranium-238 (720 p/Cil) (p. 5-4).

B: estimated, below contract required detection limits. J: estimated quantity.
K: unknown bias. μg/l = micrograms per liter. pCi/l = picocuries per liter.

Table D.8 Substances present in Pit 4 sediments at the Midnite Mine

Source	Maximum Concentrations of Pit 4 Sediment Contaminants
Superfund Technical Assessment and Response Team 1998	arsenic (25.4 mg/kg), barium (131 mg/kg), beryllium (3.6 mg/kg), chromium (30.5 mg/kg), cobalt (23.1 mg/kg), copper (26.7 mg/kg), lead (29.7 mg/kg), manganese (1,820 mg/kg), nickel (38.6 mg/kg), vanadium (46.1 mg/kg), zinc (79.8 mg/kg), uranium (772 mg/kg), radium-226 (38.050 pCi/kg), radium-228 (5,890 pCi/kg), thorium-230 (100,400 pCi/kg), thorium-232 (1,450 pCi/kg), uranium-234 (312,000 pCi/kg), uranium-235 (14,800 JK pCi/kg), uranium-238 (236,800 pCi/kg) (p. 5-4, p. 5-5).

J: estimated quantity. K: unknown bias. mg/kg = milligrams per kilograms.
pCi/kg = picocuries per kilogram.

Table D.9 Substances present in Pollution Control Pond water at the Midnite Mine

Data Source	PCP Water Characteristics and Contaminants
Sumioka 1991	pH 3.5-3.7which is below EPA standards for aquatic life and uranium mine effluents (pp. 32-33); aluminum, beryllium, cadmium, calcium, copper, iron, magnesium, manganese, sulfate and zinc exceed EPA aquatic life criteria (p. 32, p. 38); radium-226 (48-63 pCi/l) and uranium (160,000-180,000 µg/l) exceed maximum for uranium mine effluents (p. 34, p. 38).
Superfund Technical Assessment and Response Team 1998	Maximum concentrations: arsenic (17.5 µg/l), barium (6.1 JBK µg/l), beryllium (53.4 µg/l), cadmium (53.4 µg/l), chromium (32.6 JK µg/l), cobalt (1,330 µg/l), copper (343 µg/l), lead (31 µg/l), manganese (115,000 µg/l), nickel (2,130 µg/l), selenium (71.7 µg/l), zinc (4,640 µg/l), uranium (26,700 µg/l), radium-226 (29.2 pCi/l), radium-228 (5.3 pCi/l), thorium-230 (791 pCi/l), thorium-232 (112 JK pCi/l), uranium-234 (5,610 pCi/l), uranium-235 (128 pCi/l), uranium-238 (6,350 pCi/l) (p. 5-5).
URS Corp 2000	Waterfowl observed on the PCP (p. 3-20).

B: estimated, below contract required detection limits. J: estimated quantity.
K: unknown bias. µg/l = micrograms per liter. pCi/l = picocuries per liter.

**Table D.10 Substances present in Pollution Control Pond sediments at the
Midnite Mine**

Data Source	Maximum Concentration of PCP Sediment Contaminants
Superfund Technical Assessment and Response Team 1998	arsenic (25.6 mg/kg), barium (689 mg/kg), beryllium (29.8 mg/kg), cadmium (11.2 mg/kg), chromium (30.4 mg/kg), cobalt (166 mg/kg), copper (751 mg/kg), lead (25.4 mg/kg), manganese (4,330 mg/kg), nickel (757 mg/kg), vanadium (16.3 mg/kg), zinc (995 JK mg/kg), uranium (5,780 mg/kg), radium-226 (5,919 pCi/kg), radium-228 (16,940 pCi/kg), thorium-228 (5,130 pCi/kg), thorium-230 (2,542 pCi/kg), thorium-232 (47,200 pCi/kg), uranium-234 (2,408,000 pCi/kg), uranium-235 (115,200 JK pCi/kg), uranium-238 (2,336,000 pCi/kg) (p. 5-5, p. 5-6).

mg/kg = milligrams per kilograms. pCi/kg = picocuries per kilogram.

Table D.11 Substances present in Blood Pool water at the Midnite Mine

Data Source	Blood Pool Water Characteristics and Contaminants
Marcy et al. 1994	Low pH and elevated levels of aluminum, calcium, iron, manganese, magnesium, nickel, sulfate, zinc give liquid blood red color (p. 16)
Superfund Technical Assessment and Response Team 1998	Maximum concentrations: arsenic (7.1 JBK µg/l), barium (5.3 JBK µg/l), beryllium (27.8 µg/l), cadmium (8.4 µg/l), chromium (27.4 JK µg/l), cobalt (792 µg/l), copper (1,190 µg/l), lead (12.5 µg/l), maganese (36,800 µg/l), nickel (1,260 µg/l), selenium (27.2 µg/l), zinc (1,160 µg/l), uranium (8,170 µg/l), radium-226 (7.3 pCi/l), radium-228 (2.4 pCi/l), thorium-230 (99.9 pCi/l), uranium-234 (2,386 pCi/l), uranium-235 (440 pCi/l), uranium-238 (2,670 pCi/l) (p. 5-6).
URS Corporation 2000	Lack of surface water connections between Blood Pool and other water ways but "URS personnel observed numerous frogs jumping into and out of the Blood Pool", fence doesn't completely surround the Blood Pool (p. 3-20).

B: estimated, below contract required detection limits. J: estimated quantity.
K: unknown bias. mg/kg = milligrams per kilograms. µg/l = micrograms per liter.
pCi/kg = picocuries per kilogram.

Table D.12 Substances present in Blood Pool sediments at the Midnite Mine

Data Source	Maximum Concentrations of Blood Pool Sediment Contaminants
Superfund Technical Assessment and Response Team 1998	antimony (1.6 JBK mg/kg), arsenic (75 mg/kg), barium (131 mg/kg), beryllium (1.1 JBK mg/kg), chromium (50.7 mg/kg), cobalt (11.3 JBK mg/kg), copper (87.8 mg/kg), lead (21.1mg/kg), maganese (352 mg/kg), nickel (29 mg/kg), vanadium (66.3 mg/kg), zinc (74.55 JK mg/kg), uranium (104 mg/kg), radium-226 (35,200 pCi/kg), radium-228 (5,420 pCi/kg), thorium-228 (1,870 pCi/kg), thorium-230 (46,000 pCi/kg), thorium-232 (1,900 pCi/kg), uranium-234 (42,500 pCi/kg), uranium-238 (32,100 pCi/kg) (p. 5-6,p. 5-7).

B: estimated, below contract required detection limits. J: estimated quantity.
K: unknown bias. mg/kg = milligrams per kilograms. pCi/kg = picocuries per kilogram.

Living in a Contaminated World

Table D.13 Substances present in drainage waters at the Midnite Mine

Location	Maximum Concentration of Seep Contaminants
West Drainage (3,000 feet)	beryllium (10 µg/l), cadmium (7.7µg/l), cobalt (190 µg/l), nickel (458 µg/l), manganese (23,400 µg/l), selenium (16.8 µg/l), zinc (798 µg/l), uranium (36,800 µg/l), radium-226 (7.6 pCi/l), thorium-232 (0.4 pCi/l), uranium-234 (1,260 pCi/l), uranium-235 (118.3 pCi/l), uranium-238 (1,502 pCi/l) (p. 6-6).
Central Drainage (1,900 feet)	beryllium (12.4 µg/l), cadmium (47.3 µg/l), cobalt (749 µg/l), lead (34.7µg/l), nickel (1,240 µg/l), magnanese (116,000 µg/l), selenium (70 µg/l), zinc (2,270 µg/l), uranium (30,600 µg/l), radium-226 (30.4 pCi/l), thorium-230 (47.3 pCi/l), uranium-234 (6,290 pCi/l), uranium-235 (457 pCi/l), uranium-238 (6,390 pCi/l) (p. 6-7).
East Drainage (6,800 feet/ 1.29 miles)	arsenic (73 µg/l), barium (916 µg/l), uranium (68.6 µg/l), radium-226 (2.2 pCi/l), uranium-234 (26.9 pCi/l), uranium-235 (2.3 pCi/l), uranium-238 (26.5 pCi/l) (p. 6-9).
550 feet Downstream of Drainage Seeps	All "dry for field work" (p. 6-2).

µg/l = micrograms per liter. pCi/l = picocuries per liter.

Source: Superfund Technical Assessment and Response Team, 1998.

Table D.14 Substances present in drainage sediments at the Midnite Mine

Location	Maximum Concentration of Sediment Contaminants
West Drainage (3,000 feet)	arsenic (4.1 mg/kg), copper (18.3 mg/kg), uranium (107 mg/kg), uranium-238 (34,470 pCi/kg) (p. 6-6).
550 feet Downstream of West Drainage	arsenic (5.8 mg/kg), copper (12.4 mg/kg), manganese (533 mg/kg), vanadium (12,8 mg/kg) (p. 6-7).
Central Drainage (1,900 feet)	arsenic (76.2 mg/kg), barium (162 mg/kg), beryllium (24.2 mg/kg), cadmium (10.2 mg/kg), chromium (33.1 mg/kg), cobalt (382 mg/kg), copper (171 mg/kg), lead (49.4 mg/kg), manganese (6,910 mg/kg), nickel (502 mg/kg), selenium (6.2 mg/kg), vanadium (14.2 mg/kg), zinc (1,080 mg/kg), uranium (4,140 mg/kg), radium-226 (262,700 pCi/kg), uranium-234 (1,432,000 pCi/kg), uranium-235 (136,500 JK pCi/kg), uranium-238 (1,432,000 pCi/kg) (p. 6-7, p. 6-8).
550 feet Downstream of Central Drainage Seep	arsenic (14.1 mg/kg), barium (119 mg/kg), beryllium (4.2 mg/kg), cadmium (2.8 mg/kg), chromium (25 mg/kg), cobalt (142 mg/kg), copper (45.2 mg/kg), manganese (5,200 mg/kg), nickel (230 mg/kg), selenium (5.2 mg/kg), zinc (399 mg/kg), vanadium (29.9 mg/kg), uranium (3,640 mg/kg), thorium-230 (20,200 pCi/kg), uranium-238 (1,106,000 pCi/kg (p. 6-8).
East Drainage (6,800 feet/ 1.29 miles)	arsenic (15.3 mg/kg), barium (100 mg/kg), chromium (24.7 mg/kg), copper (29.1 mg/kg), manganese (889 mg/kg), nickel (17.8 mg/kg), vanadium (51.5 mg/kg), zinc (63.5 mg/kg), uranium (83 µg/l), radium-226 (12,390 pCi/kg), thorium-230 (32,700 pCi/kg), uranium-234 (16,410 pCi/kg), uranium-238 (19,350 pCi/kg) (p. 6-9, p. 6-10).

J: estimated quantity. K: unknown bias. mg/kg = milligrams per kilograms.
pCi/kg = picocuries per kilogram.

Source: Superfund Technical Assessment and Response Team Final ESI Report, 1998.

Table D.14 Continued

550 feet Downstream of East Drainage Seep	arsenic (117.8 mg/kg), barium (254 mg/kg), beryllium (4.2 mg/kg), cadmium (9.6 mg/kg), chromium (27 mg/kg), cobalt (35.9 mg/kg), manganese (24,300 mg/kg), selenium (16 mg/kg), vanadium (35.5 mg/kg), zinc (395 JK mg/kg), uranium (78.5 mg/kg), throium-230 (34,500 pCi/kg), uranium-234 (13.3 pCi/kg), uranium-235 (0.8 pCi/kg), uranium-238 (29,680 pCi/kg) (p. 6-10, p. 6-11).

J: estimated quantity. K: unknown bias. mg/kg = milligrams per kilograms.
pCi/kg = picocuries per kilogram.

Source: Superfund Technical Assessment and Response Team Final ESI Report, 1998.

Table D.15 Substances present in other waterways at the Midnite Mine

Data Source	Other Water Contaminants
Superfund Technical Assessment and Response Team 1998	Springs, seeps and streams: Radium-226, Uranium-234, Uranium 238 'were generally elevated above regional levels with maximum concentration detected up to 500pCi/l' (p. 5-7).
	NPDES Outfall: barium (78.1 JBK µg/l), chromium (246 JBK µg/l), copper (8.4 JBK µg/l), manganese (369 µg/l), zinc (5.2 JBK µg/l), uranium (82 µg/l), uranium-234 (13.2 pCi/l), uranium-235 (0.7 pCi/l), uranium-238 (12.8 JH pCi/l); none exceed daily max or average discharge limits (p. 5-7).
	NPDES Outfall sediment concentrations: antimony (1.8 JBK mg/kg), arsenic (82.1 mg/kg), barium (112 mg/kg),cadmium (0.75 JBK mg/kg), chromium (39.2 mg/kg), cobalt (23.1 mg/kg), copper (57.5 mg/kg), lead (32.7 mg/kg), manganese (3,090 mg/kg), nickel (37.1 mg/kg), vanadium (32.8 mg/kg), zinc (142 JK mg/kg), uranium (406 mg/kg), radium-228 (5,100 pCi/kg), thorium-230 (50,300 pCi/kg), thorium-232 (4,520 pCi/kg), uranium-234 (126,100 JK pCi/kg), uranium-235 (5,410 JK pCi/kg), uranium-238 (136,600 pCi/kg) (p. 5-7).
Shepherd Miller, Inc 1991	Effluent limits not exceeded except for manganese (p. 18); encountered a deep bedrock aquifer in Pit 3 (p. 14).
Marcy 1993	Boyd seep on east side of mine: acidic water source with elevated toxic metal concentrations (p. 476).
Marcy et al. 1994	Boyd seep on east side of mine: acidic water with high concentrations of calcium, magnesium and sulfate (p. 16).
	Drainage from protore piles: highest sulfate concentration on mine site (p. 20).

B: estimated, below contract required detection limits. J: estimated quantity.
K: unknown bias. µg/l = micrograms per liter. pCi/l = picocuries per liter.
mg/kg = milligrams per kilograms. pCi/kg = picocuries per kilogram.

Living in a Contaminated World

Table D.16 Other substances of concern present at the DMC mill site

Substance	ug/l or ppm in soil	Proposed clean-up level		Back-ground
		Residential	Industrial	
Arsenic	20-200	20	200	<2.5
Cadmium	0.83-2.32	2	10	<0.17
Lead	62-1100	250	1000	4.2
Mercury	0.76-12.7	1	1	<0.10
Copper	32-160	na	na	7.2
Manganese	492-1930	na	na	113

μg/l = micrograms per liter. ppm = parts per million. na = not applicable.

Source: Washington State Department of Health, 1991.

Appendix E

Maps and Photos of Teledyne Wah Chang Albany Operations and Area

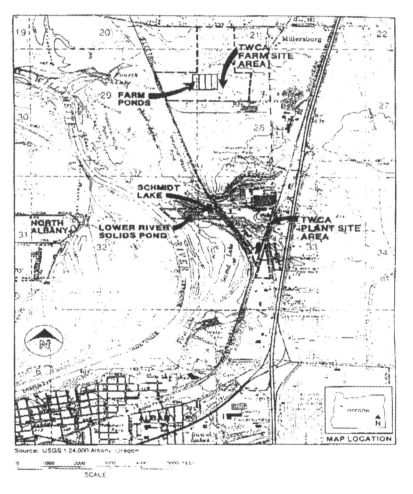

Figure E.1   Map of Teledyne Wah Chang Superfund Site and surrounding area

Source: Environmental Protection Agency 1989, 1994

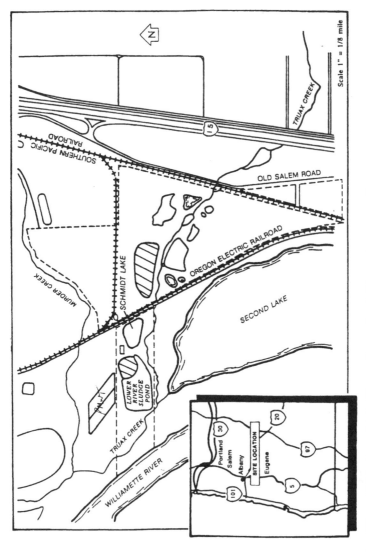

Figure E.2   Map of Teledyne Wah Chang Superfund Site and surrounding area
Source: Environmental Protection Agency Administrative Record for *Teledyne Wah Chang Albany*,
Item Number 11.1.0000001, Seattle WA.

Figure E.3   Aerial view of Millersburg, Oregon
Source: Albany-Millersburg Economic Development Council.

Appendix F

Maps and Photos of Dawn Mining Company Operations and Area

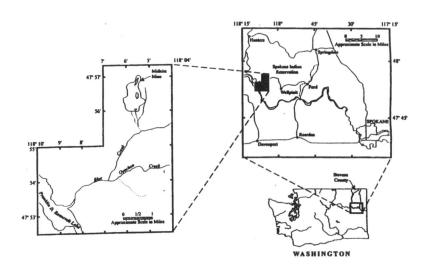

Figure F.1   Map of Dawn Mining Company's Midnite Mine, mill site and surrounding area

Source: Washington State Department of Social and Health Services, 1989.

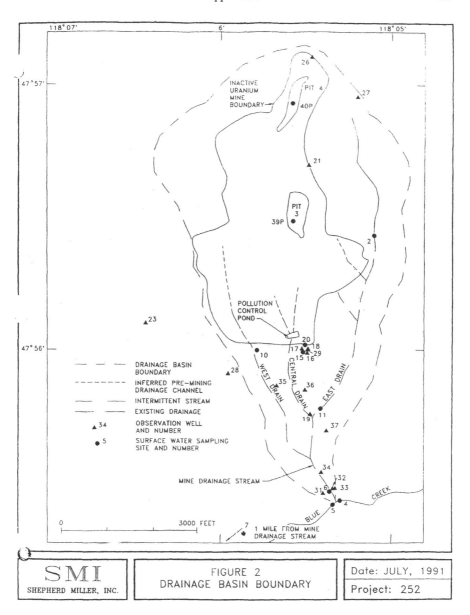

Figure F.2 Map of Midnite Mine drainages
Source: Shepherd Miller, Inc., 1991.

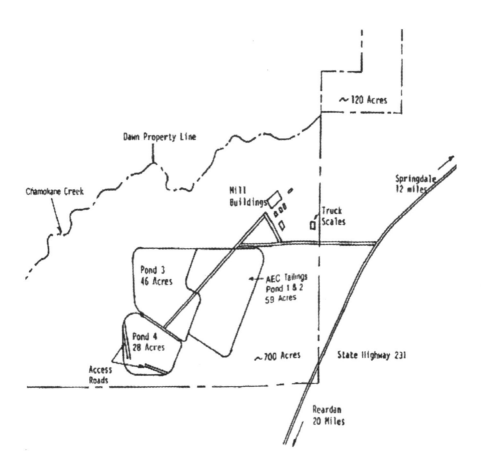

Figure F.3 Map of Dawn Mining Company's mill site
Source: Washington Department of Social and Health Services, 1980.

Figure F.4 The town of Ford, Washington
Source: Ellen Omohundro, 2002.

Figure F.5 The old school, Ford, Washington
Source: Ellen Omohundro, 2002

Figure F.6 The area near the Dawn Mining Company mill site, Washington
Source: Ellen Omohundro, 2002

Figure F.7 Midnite Mine, Pit 3
Source: Environmental Protection Agency,
http://yosemite.epa.gov/r10/cleanup.nsf/9f3c21896330b4898825687b007a0f33/25f579940d
8b88256744000327a5?OpenDocument

Figure F.8 Midnite Mine, Pit 3
Source: Environmental Protection Agency,
http://yosemite.epa.gov/r10/cleanup.nsf/9f3c21896330b4898825687b007a0f33/25f579940d
8b88256744000327a5?OpenDocument

Figure F.9 Midnite Mine entrance
Source: Environmental Protection Agency,
http://yosemite.epa.gov/r10/cleanup.nsf/9f3c21896330b4898825687b007a0f33/25f579940d
8b88256744000327a5?OpenDocument

Bibliography

Alasuutari, P. (1995), *Researching Culture: Qualitative Method and Cultural Studies*, Sage Publications, Thousand Oaks, CA.

Aldis, H. (23 November 1981), *Status Report - Teledyne*, Ecology and Environment, Inc., Reference No.TDD 10-8005-08, Environmental Protection Agency, Site Listing Docket for the Teledyne Wah Chang Albany Superfund Site, Seattle, WA.

Allen, H.E. (8 February 1999), *Persistent, Bioaccumulative, and Toxic (PBT) Chemicals: Considerations for RCRA Waste Minimization of Metals*, EPA Docket No. F-98-MMLP-FFFF, Environmental Protection Agency, Site Listing Docket for the Teledyne Wah Chang Albany Superfund Site, Seattle, WA.

Allen, J.C. and Dillman, D.A. (1994), *Against all Odds: Rural Community in the Information Age*, Westview Press, Boulder, CO.

Alley, B.F. (1889), *Linn County Oregon, Descriptive and Resources, Its Cities and Towns*, Boyce and Hibler Electric Power Print, Albany, OR.

American National Library (October 2001), 'Natural Decay Series: Uranium, Radium, and Thorium', found at http://www.oversight.state.id.us/ov_library/Contaminant_Fact_Sheets/DecaySeries_FactSheet_ANL.pdf, 28 March 2003.

Anderson, A. (1997), *Media, Culture and the Environment*, Rutgers University Press, New Brunswick, NJ.

Axelrod, L.J. (1994), 'Balancing Personal Needs with Environmental Preservation: Identifying the Values that Guide Decisions in Ecological Dilemmas', *Journal of Social Issues*, Vol. 50, pp. 85-104.

Baker, D. and Banks, B.D. (16 February 1999), *Comments of Newmont Gold Company on the United States Environmental Protection Agency's Proposed Addition of the Midnite Mine site to the National Priorities List, 64 Federal Regulation 7564*, Environmental Protection Agency, Site Listing Docket for the Midnite Mine Superfund Site, Seattle, WA.

Beamish, T.D. (2002), *Silent Spill: The Organization of an Industrial Crisis*, MIT Press, Cambridge, MA.

Beck, U. (1992), 'From Industrial Society to the Risk Society: Questions of Survival, Social Structure and Ecological Enlightenment', *Theory, Culture and Society*, Vol. 9, pp. 97-123.

Beck, U. (1995), *Ecological Enlightenment: Essays on the Politics of the Risk Society*, Humanities Press, Atlantic Heights, NJ.

Been, V. (1994), 'Locally Undesirable Land Uses in Minority Neighborhoods: Disproportionate Siting or Market Dynamics?', *The Yale Law Journal*, Vol. 103, pp. 1383-1422.

Bell, C. and Newby, H. (1972), *Community Studies: An Introduction to the Study the Local Community*, Praeger, New York.

Bellrose, C.A. and Pilisuk, M. (1991), 'Vocational Risk Tolerance and Perceptions of Occupational Hazards', *Basic and Applied Social Psychology*, Vol. 12, pp. 303-323.

Betz. J. (30 August 1982), *Potential Hazardous Waste Site, Environmental Protection Agency Site Inspection Report*, Environmental Protection Agency, Site Listing Docket for the Teledyne Wah Chang Albany Superfund Site, Seattle, WA.

Bohm, F.C. and Craig, E.H. (1983), *The People's History of Stevens County*, The Stevens County Historical Society, Colville, WA.

Boholm, A. (1998), 'Comparative Studies of Risk Perception: A Review of Twenty Years of Research', *Journal of Risk Research*, Vol. 1, pp. 135-163.

Bowen, W.A. (1978), *The Willamette Valley: Migration and Settlement on the Oregon Frontier*, University of Washington Press, Seattle, WA.

Bridger, J.C. and Luloff, A.E. (1998), 'Sustainable Community Development: An Intersectional Perspective', in E. Zuber, S. Nelson, and A.E. Luloff (eds) *Community: A Biography in Honor of the Life and Work of Ken Wilkinson*, Northwest Regional Center for Rural Development, State College, PA, available at http://www.cas.nercrd.psu.edu/Community/ Legacy/ bridger _community.htm.

Brown, P. (1992), 'Popular Epidemiology and Toxic Waste Contamination: Lay and Professional Ways of Knowing', *Journal of Health and Social Behavior*, Vol. 33, pp. 267-281.

Brown, P., Kroll-Smith, S. and Gunter, V.J. (2000), 'Knowledge, Citizens, and Organizations: An Overview of Environments, Diseases, and Social Conflict', in S. Kroll-Smith, P. Brown, and V. J. Gunter (eds), *Illness and the Environment: A Reader in Contested Medicine*, New York University Press, New York, pp. 9-28.

Brown, R.B., Geertsen, H.R, and Krannich, R.S. (1989), 'Community Satisfaction and Social Integration in a Boomtown: A Longitudinal Analysis', *Rural Sociology*, Vol. 54, pp. 568-586.

Bryant, B. (ed.) (1995), *Environmental Justice: Issues, Polities, and Solutions*, Island Press, Washington, DC.

Buckner, J.C. (1988), 'The Development of an Instrument to Measure Neighborhood Cohesion', *American Journal of Community Psychology*, Vol. 16, pp. 711-791.

Bullard, R.D. (1990), *Dumping in Dixie: Race, Class and Environmental Quality*, Westview Press, Boulder, CO.

Bunker Hill Superfund Task Force (1994), *The Bunker Hill Superfund Project: Successes and Frustrations*, Pan Handle Health District, Kellogg, ID.

Bureau of Indian Affairs (1870), *Report on the Commission of Indian Affairs for the Territories of Washington and Idaho, and the State of Oregon for the Year 1870*, reprinted in 1981 by YE Galleon Press, Fairfield, WA.

Bureau of Land Management (26 January 1996a), 'Notice of Intent to Prepare an Environmental Impact Statement', *Federal Register*, Vol. 61(18), p. 2528.

Bureau of Land Management (1996b), *Scoping Summary Report for Midnite Uranium Mine Reclamation*, BLM, Spokane District Office, Spokane, WA.

Bureau of Land Management (1996c), *EIS Data Needs Report for Midnite Uranium Mine Reclamation*, BLM, Spokane District Office, Spokane, WA.

Cairns, J., Jr. and Niederlehner, B.R. (1996), 'Developing a Field of Landscape Ecotoxicology', *Ecological Applications*, Vol. 6, pp. 790-796.

Cardin, D.J., Lappert, M.F. and Raston, C.L. (1986), *Chemistry of Organo-Zirconium and Hafnium Compounds*, Halsted Press, New York.

Carollo, R. (23 October 1987), 'Neighbors Reject Tainted Dirt', *Spokesman Review*, Spokane, WA.

Carr, D.S. and Halvorsen, K. (2001), 'An Evaluation of Three Democratic, Community-based Approaches to Citizen Participation: Surveys, Conversations with Community Groups, and Community Dinners', *Society and Natural Resources*, Vol. 14, pp. 107-126.

Carroll, M.S. (1995), *'Community and the Northwestern Logger: Continuities and Changes in the Era of the Spotted Owl'*, Westview Press, Boulder, CO.

Carson, R.L. (1962), *Silent Spring*, Houghton Mifflin Company, Boston.

Center for Hazardous Waste Management, Illinois Institute of Technology/IIT Research Institute (1989), *Coalition on Superfund Research Report: Consolidated Research Report, Volumes 1-2*, Coalition on Superfund, Washington, D.C.

Chemical Rubber Company (2002), *CRC Handbook of Chemistry and Physics,* CRC Press, Cleveland, Ohio.

Chess, C., Salomone, K.L. and Hance, B.J. (1995), 'Improving Risk Communication in Government: Research Priorities', *Risk Analysis,* Vol. 15, pp. 127-135.

Clark, R.C. (1927), *History of the Willamette Valley Oregon,* The S.J. Clarke Publishing Company, Chicago.

Clarke, L. (1989), *Acceptable Risk? Making Decisions in a Toxic Environment*, University of California Press, Berkeley, CA.

Clarke, L. and Short, J.F., Jr. (1993), 'Social Organization and Risk: Some Current Controversies', *Annual Review of Sociology,* Vol. 19, pp. 375-399.

Cleckley, E.W. (1997), 'The Missing Link: Effects on Communities, Neighborhoods, and Individuals', in R. Clark and L. Canter (eds), *Environmental Policy and NEPA: Past, Present, and Future,* St. Lucie Press, Boca Raton, FL, pp. 251-260.

Cohen, A.P. (1985), *The Symbolic Construction of Community*, Ellis Horwood Limited, Chichester, England.

Coleman, J.S. (1957), *Community Conflict,* The Free Press, New York.

Confederated Tribes of the Grand Ronde Community Cultural Resources Division (2002), 'Tribal History', found at http://www.grandronde.org/cultural/index.html, 23 July 2002.

Cook, S.F. (1955), *The Epidemic of 1830-33 in California and Oregon,* University of California Press, Berkely, CA.

Cottrell, L.S., Jr. (1977), 'The Competent Community', in R.L. Warren (ed.), *New Perspectives on the American Community* (third edition), Rand McNally College Publishing Company, Chicago, pp. 546-560.

Couch, S.R. and Kroll-Smith, S. (1990), 'Patterns of Victimization and Chronic Technological Disaster', in E.C. Viano (ed.), *The Victimology Handbook: Research Findings, Treatment, and Public Policy,* Garland Publishing, Inc., New York, pp. 159-176.

Couch, S.R. and Kroll-Smith, S. (1994), 'Environmental Controversies, Interactional Resources, and Rural Communities: Siting Versus Exposure Disputes', *Rural Sociology,* Vol. 59, pp. 25-44.

Cuba, L and Hummon, D.M. (1993), 'A Place to Call Home: Identification with Dwelling, Community and Region', *The Sociological Quarterly,* Vol. 34, pp. 111-131.

Dalton, R.J., Garb, P., Lovrich, N.P., Pierce, J.C. and Whiteley, J.M. (1999), *Critical Masses: Citizens, Nuclear Weapons Production, and Environmental Destruction in the United States and Russia,* MIT Press, Cambridge, MA.

Darst, H., Enerson, R., George, D., Johnson, S., Joo, M., McDaniel, D., McManamin, N., Sullivan, K. and Walke, R. (1979), *Zirconium Hazards and Nuclear Profits: A Report on Teledyne Wah Chang Albany,* Pacific Northwest Research Center, Eugene, OR, also at Environmental Protection Agency, Site Listing Docket for the Teledyne Wah Chang Albany Superfund Site, Seattle, WA.

DeGuire, M.F. (1985), *Dawn Mining Company NPDES Permit Application Supporting Information*, Dawn Mining Company, Ford, WA, also at Environmental Protection Agency, Site Listing Docket for the Midnite Mine Superfund Site, Seattle, WA..

Department of Energy (1989), *Radioactive Decay Tables: A Handbook of Decay Data for Application to Radioactive Dosimetry and Radioactive Assessments*, Department of Energy, Washington, DC.

Dietz, T., Frey, R.S. and Rosa, E.R. (2001), 'Risk, Technology and Society', in R.E. Dunlap, and W. Michelson (eds), *Handbook of Environmental Sociology,* Greenwood, Westport, CT, pp. 329-369.

Dietz, T., Stem, P.C., and Rycroft, R.W. (1989), 'Definitions of Conflict and Legitimation of Resources: The Case of Environmental Risk', *Sociological Forum,* Vol. 4(l), pp. 47-70.

Dombrowsky, W.R. (1998), 'Again and Again: Is a Disaster What We Call a "Disaster?"', in E.L. Quarantelli (ed.) *What is a Disaster?,* Routledge, New York, pp. 19-30.

Doreian, P. and Stokman, F.N. (1997), *Evolution of Social Networks,* Gordon and Breach Publishers, Amsterdam, Netherlands.

Douglas, M. and Wildavsky, A. (1983), *Risk and Culture: An Essay on the Selection of Technological and Environmental Dangers,* University of California Press, Berkeley, CA.

Dunlap, R.E., Kraft, M.E., and Rosa, E. (eds) (1993), *Public Reactions to Nuclear Waste: Citizens' View of Repository Siting,* Duke University Press, Durham, NC.

Durkheim, E. (1933), *The Division of Labor in Society,* The Free Press, New York.

Edelstein, M.R. (1988), *Contaminated Communities: The Social and Psychological Impacts of Residential Toxic Exposures,* Westview Press, Boulder, CO.

Environmental Protection Agency (19 January 1981), 'Responsiveness Summary and Preamble on Public Participation Policy', *Federal Register,* Vol. 46(12), pp.1-30.

Environmental Protection Agency (1986), *Notice of Proposed Issuance of a National Pollutant Discharge Elimination System (NPDES) Permit to Discharge Pollutants Pursuant to the Provisions of the Clean Water Act and Notice of Water Quality Certification,* Washington, DC.

Environmental Protection Agency (November 1988), 'Preliminary Assessment Petition Fact Sheet', found at http://www.epa.gov/r10earth/offices/oec/pa_petit.pdf, 26 March 2003.

Environmental Protection Agency (1989), *Record of Decision for Teledyne Wah Chang Albany,* EPA Region 10, Seattle, WA, 90/021.

Environmental Protection Agency (November 1992), *The Hazard Ranking System Guidance Manual: Interim Final,* NTIS PB92-963377, EPA 9345.1-07, Washington, DC.

Environmental Protection Agency (15 September 1994a), *National Oil and Hazardous Substances Pollution Contingency Plan (NCP),* 40 CFR 300, EPA, Washington, DC.

Environmental Protection Agency (1994b), *Record of Decision, for Teledyne Wah Chang Albany,* EPA Region 10, Seattle, WA, 94/078.

Environmental Protection Agency (1995), *Record of Decision for Teledyne Wah Chang Albany,* EPA Region 10, Seattle, WA, 95/125.

Environmental Protection Agency (January 2000a), 'This is Superfund: A citizen's guide to EPA's Superfund Program,' EPA 540-K-99-006 OSWER 9200.5-12A 2000, found at http://www.epa.gov/superfund/whatissf/sfguide.html, 26 March 2003.

Environmental Protection Agency (28 December 2000b), 'Draft Public Involvement Policy [FRL-6923-9]', *Federal Register,* Vol. 65(250), pp. 82335-82345.

Environmental Protection Agency (24 August 2000c), 'National Priority List', *Federal Register,* Vol. 65(165), pp. 51567-51571.

Environmental Protection Agency (2000d), *Support Document for the Revised National Priorities List Final Rule May 2000,* State, Tribal and Site Identification Center, Office of Solid Waste and Emergency Response, U.S. Environmental Protection Agency, Washington, DC.

Environmental Protection Agency (2001a), 'EPA Dialogue Summary: Local Issues and Superfund', found at http://www.network-democracy.org/epa-pip/join/agenda. html, 19 July 2001.

Environmental Protection Agency (2001b), *Teledyne Wah Chang Albany Superfund Site, Explanation of Significant Differences to the September 25, 1995 Record of Decision: Final Remedial Action for Surface and Subsurface Soil Operable Unit,* EPA Region 10, Seattle, WA.

Environmental Protection Agency (4 October 2002a), 'Triumph Mine Tailings Piles, Triumph, Idaho Site Narrative ID# IDD984666024', found at http://www.epa.gov/superfund/sites/npl/nar1370.html, http://www.epa.gov/superfund/sites/query/advquery.html, 26 March 2003.

Environmental Protection Agency (April 2002b), 'Murray Smelter, Utah Site Description EPA ID # UTD 980951420', found at http://www.epa.gov/superfund/sites/npl/nar1415.html, http://www.epa.gov/superfund/sites/query/advquery.html, 26 March 2003.

Environmental Protection Agency (4 October 2002c), 'Big River Mine Tailings/St. Joe Minerals Corp. Site Description EPA ID # MOD981126899', found at http://www.epa.gov/superfund/sites/npl/nar1336.html, http://www.epa.gov/superfund/sites/query/advquery.html, 26 March 2003.

Environmental Protection Agency (July 2002d), 'Love Channel, New York Site Description EPA ID # NYD000606947', found at http://www.epa.gov/superfund/sites/npl/nar180.html, http://www.epa.gov/superfund/sites/query/advquery.html, 26 March 2003.

Environmental Protection Agency (4 October 2002e), 'Summitville Mine, Rios Grande County EPA ID # MOD980685226', found at http://www.epa.gov/superfund/sites/npl/nar833.html, http://www.epa.gov/superfund/sites/query/advquery.html, 26 March 2003.

Environmental Protection Agency (4 October 2002f), 'Times Beach, Missouri Site Description EPA ID # MOD980685226', found at http://www.epa.gov/superfund/sites/npl/nar833.html, http://www.epa.gov/superfund/sites/query/advquery.html, 26 March 2003.

Environmental Protection Agency (April 2002g), *Superfund Community Involvement Handbook,* EPA 540-K-01-003, Washington, DC.

Environmental Protection Agency (2003a), 'Comprehensive Environmental Response, Compensation, and Liability Information System (CERCLIS) Overview', found at http://yosemite.epa.gov/r10/cleanup.nsf/1887fc8b0c8f2aee8825648f00528583/6fc5afc8d542a08688256506005509c7?OpenDocument, 26 March 2003.

Environmental Protection Agency (2003b), 'Technical Assistant Grant (TAG) Program', found at http://www.epa.gov/superfund/whatissf/sfguide.html, 26 March 2003.

Erikson, K. (1994), *A New Species of Trouble: The Human Experience of Modern Disasters,* W. W. Norton and Company, New York.

Etzioni, A. (1993), *The Spirit of Community: Rights, Responsibilities, and the Communitarian Agenda,* Crown Publishers, New York.

Etzioni, A. (1996), *The New Golden Rule: Community and Morality in a Democratic Society,* Basic Books, New York.

Everest, L. (1986), *Behind the Poison Cloud: Union Carbide's Bhopal Massacre,* Banner Press, Chicago.

Eyles, J. (1997), 'Environmental Health Research: Setting an Agenda by Spinning Our Wheels or Climbing the Mountain?', *Health and Place,* Vol. 3(1), pp. 1-13.

Filkins, R., Allen, J.C. and Cordes, S. (2000), 'Predicting Community Satisfaction Among Rural Residents: An Integrative Model', *Rural Sociology,* Vol. 65, pp. 72-86.

Finkel, A.M. and Golding, D. (1994), *Worst Things First? The Debate Over Risk-based National Environmental Priorities,* Resources for the Future, Washington, DC.

Flett, B. (1997), 'Interviews with Elders', found at http://www.spokanetribe.com/voices.html, 25 June 2002.

Flynn, J., Peters, E., Mertz, C.K. and Slovic, P. (1998), 'Risk, Media and Stigma at Rocky Flats', *Risk Analysis*, Vol. 18, pp. 715-727.

Flynn, J., Slovic, P. and Mertz, C.K. (1994), 'Gender, Race, and Perception of Environmental Health Risks', *Risk Analysis*, Vol. 14, pp. 1101-1108.

Fogleman, V.M. (1990), *Guide to the National Environmental Policy Act: Interpretation, Applications, and Compliance*, Quorum Books, Westport, CT.

Forman, R.T.T. (1995), *Land Mosaics: The Ecology of Landscapes and Regions*, Cambridge University Press, New York.

Fowlkes, M.R. and Miller, P.Y. (1982), *Love Canal: The Social Construction of Disaster*, Federal Emergency Management Agency, Washington, DC.

Freese, L. (1997), *Environmental Connections*, JAI Press, Inc., Greenwich, CT.

Freie, J.F. (1998), *Counterfeit Community: The Exploitation of Our Longing for Connectedness*, Rowman and Littlefield Publishers, Inc., New York.

Freudenburg, W.R. (1992), 'Public Responses to Technological Risks: Toward a Sociological Perspective', *The Sociological Quarterly*, Vol. 33(3), pp. 389-412.

Frey, J.H. (1993), 'Risk Perceptions Associated with a High-level Nuclear Waste Repository', *Sociological Spectrum*, Vol. 13, pp. 139-151.

Gans, H.J. (1962), *The Urban Villagers: Group and in the Life of Italian-Americans*, The Free Press, New York.

Gilbert, C. (1998), 'Studying Disaster: Changes in the Main Conceptual Tools', in E.L. Quarantelli (ed.), *What is a Disaster?*, Routledge, New York, pp. 11-18.

Goffman, E. (1963), *Stigma: Notes on the Management of Spoiled Identity*, Simon and Schuster, Inc., New York.

Goodsell, T. (2000), 'Maintaining Solidarity: A Look Back at the Mormon Village', *Rural Sociology*, Vol. 65, pp. 357-375.

Goszczynska, M., Tyszkz, T. and Slovic, P. (1991), 'Risk Perception in Poland: A Comparison with Three Other Countries', *Journal of Behavioral Decision Making*, Vol. 4, pp. 179-193.

Gould, C.C. (1978), *Marx's Social Ontology: Individuality and Community in Marx's Theory of Social Reality*, The MIT Press, Cambridge, MA.

Gray, B. (2003), 'Framing of Environmental Disputes', in R.J. Lewicki, B. Gray, and M. Elliot (eds), *Making Sense of Intractable Environmental Conflicts: Concepts and Cases*, Island Press, Washington, DC, pp. 11-34.

Halvorsen, K.E. (2001), 'Relationships Between National Forest System Employee Diversity and Beliefs Regarding External Interest Groups', *Forest Science*, Vol. 47, pp. 258-269.

Halvorsen, K.E. (2003), 'Assessing the Effects of Public Participation', *Public Administration Review*, Vol. 63, 535-543.

Hane, F.J. (September 1989), 'Early Uranium Mining in the United States', paper presented at *Uranium Institute of London, Fourteenth International Symposium*, found at http://www.world-nuclear.org/usumin.htm, 13 June 2003.

Hannigan, J.A. (1995), *Environmental Sociology: A Social Constructionist Perspective*, Routledge, New York.

Hanson, R.D. (2001), 'Half Lives of Reagan's Indian Policy: Marketing Nuclear Wastes to American Indians', *American Indian Culture and Research Journal*, Vol. 25, pp. 21-44.

Harmon, G. D. (1941), *Sixty Years of Indian Affairs: Political, Economic, and Diplomatic 1789-1850*, The University of North Carolina Press, Chapel Hill, NC.

Harper, B.L., Flett, B., Harris, S., Abeyta, C. and Kirschner, F. (2002), 'The Spokane Tribe's Multipathway Subsistence Exposure Scenario and Screening Level RME', *Risk Analysis* Vol. 22(3), pp. 513-525.

Hedrick, J.B. (1 October 2002), 'Zirconium and Hafnium', *Minerals Yearbook 2001*, U.S. Geological Survey Publications, found at http://minerals.usgs.gov/minerals/pubs/pubs/commodity/zirconium/730401.pdf, 13 June 2003.

Hewitt, K. (1998), 'Excluded Perspectives in the Social Construction of Disaster', in E.L. Quarantelli (ed.), *What is a disaster?*, Routledge, New York, pp. 75-91.

Hillery, G. (1955), 'Definitions of Community: Areas of Agreement', *Rural Sociology*, Vol. 20, p. 118.

Hird, J.A. (1993), 'Environmental Policy and Equity: The Case of Superfund', *Journal of Policy Analysis and Management*, Vol. 12, pp. 323-343.

Howard, R.E. (1995), *Human Rights and the Search for Community*, Westview Press, Boulder, CO.

Ichniowski, T. (12 July 2001), 'Study Pegs Superfund's Cost at $14-16 Billion Through 2009', *Engineering News-Record*, found at http://www.construction.com/News Center/Headlines/ ENR/20010712d.jsp.

Imbroscio, D.L. (1997), *Reconstructing City Politics: Alternative Economic Development and Urban Regimes*, Sage Publications, Thousand Oaks, CA.

Immigration and Naturalization Service (2002), 'This Month in Immigration History: March 1790', found at http://www.ins.usdoj.gov/graphics/aboutins/history/mar1790. htm, 23 July 2003.

Interorganizational Committee on Guidelines and Principles for Social Impact Assessment (1997), 'Putting People in the Environment: Principles for Social Impact Assessment', in R. Clark and L. Canter (eds), *Environmental Policy and NEPA: Past, Present, and Future,* St. Lucie Press, Boca Raton, FL, pp. 229-250.

Jaeger, C.C., Renn, O., Rosa, E.A. and Webler, T. (2001), *Risk, Uncertainty, and Rational Action*, Earthscan Publications Ltd., London.

Jenkins, R. (1996), *Social Identity,* Routledge, New York.

Jessett, T.E. (1960), *Chief Spokan Garry 1811-1892: Christian, Statesman, and Friend of the White Man*, T.S. Denison and Company, Inc., Mineapolis, MN.

Jobes, P.C., Stinner, W.F. and Wardwell, J.M. (1992), *Community, Society and Migration: Noneconomic Migration in America,* University Press of America, New York.

Johnson, S.V. (1904), *A Short History of Oregon*, A.C. McClurg and Co., Chicago.

Jones, E.E., Farina, A., Hastorf, A.H., Markus, H., Miller, D.T., Scott, R.A. and French, R de S. (1984), *Social Stigma: The Psychology of Marked Relationship*, W.H. Freeman, New York.

Kasperson, R.E., Renn, O., Slovic, P., Brown, H.S., Emel, J., Goble, R., Kasperson, J.X. and Ratick, S. (1988), 'The Social Amplification of Risk: A Conceptual Framework', *Risk Analysis*, Vol. 8, pp. 177-187.

Kaufman, H.F. (1977), 'Toward an Interactional Conception of Community', in R.L. Warren (ed.), *New Perspectives on the American Community* (third edition*)*, Rand McNally College Publishing Company, Chicago, pp. 75-90.

Kemmis, D. (1990), *Community and the Politics of Place*, University of Oklahoma Press, London.

Kemmis, D. (1995), *The Good City and the Good Life,* Houghton Mifflin Company, New York.

Kincheloe, J.W. (16 August 1978), Memo to Mr. Peter W. McSwain, Hearing Officer, Oregon Department of Environmental Quality, regarding Application OR-100111-2, Willamette River via Truax Creek from Teledyne Wah Chang Albany, Environmental Protection Agency, Site Listing Docket for the Teledyne Wah Chang Albany Superfund Site, Seattle, WA.

Kirkpatrick, F.G. (1986), *Community: A Trinity of Models*, Georgetown University Press, Washington, DC.

Klaassen, C.D., Amdur, M.O. and Doull J. (eds) (1986), *Casarett and Doull's Toxicology: The Basic Science of Poisons* (third edition), Macmillan Publishing Company, New York.

Kleinhesselink, R.R. and Rosa, E.A. (1991), 'Cognitive Representation of Risk Perceptions: A Comparison of Japan and the United States', *Journal of Cross-Cultural Psychology*, Vol. 22, pp. 11-28.

Komter, A. (2000), 'Editorial Perspectives on Solidarity', *The Netherlands Journal of Social Sciences,* Vol. 35, pp. 3-14.

Krimsky, S. and Golding, D. (eds) (1992), *Social Theories of Risk*, Praeger, Westport, CT.

Kroll-Smith J.S. and Couch, S.R. (1991), 'As if Exposure to Toxins Were Not Enough: The Social and Cultural System as a Secondary Stressor', *Environmental Health Perspectives*, Vol. 95, pp. 61-66.

Kuletz, V. (1998), *The Tainted Desert: Environmental and Social Ruin in the American West,* Routledge, New York.

Kunreuther, H. and Slovic, P. (1999), 'Coping with Stigma: Challenges and Opportunities', *Risk: Health, Safety and Environment*, Vol. 10, pp. 269-280.

Lacy, W.B. (2000), 'Empowering Communities Through Public Work, Science, and Local Food Systems: Revisiting Democracy and Globalization', *Rural Sociology*, Vol. 65, pp. 3-26.

Landy, M.C., Roberts, M.J., and Thomas, S.R. (1994), *The Environmental Protection Agency: Asking the Wrong Questions from Nixon to Clinton,* Oxford University Press, New York.

Levine, A.G. and Stone, R.A. (1986), 'Threats to People and What They Value: Residents Perceptions of the Hazards of Love Canal', in A.H. Lebovitis, A. Baum and J.E. Singer (eds), *Advances in Environmental Psychology Volume 6 Exposure to Hazardous Substances,* Lawrence Erlbaum Associates, Hillsdale, NJ, pp. 109-130.

Linn County Pioneer Memorial Association (1979), *History of Linn County,* Work Projects Administration, State of Oregon, Eugene, OR.

Lynch, K. (1988), 'A Place Utopia', in R.L. Warren and L. Lyon (eds), *New Perspectives on the American Community* (fifth edition), The Dorsey Press, Chicago, pp. 432-443.

Lynd, R.S. and Lynd, H.M. (1929), *Middletown: A Study in Contemporary American Culture,* Harcourt, Brace Jovanovich, New York.

Lynd, R.S. and Lynd, H.M. (1937), *Middletown in Transition*, Harcourt, Brace and Jovanovich, New York.

Machlis, G. and Rosa, E.A. (1990), 'Desired Risk: Broadening the Social Amplification of Risk Framework', *Risk Analysis*, Vol. 10, pp. 161-168.

Mackey, H. (1974), *The Kalapuyans: A Sourcebook on the Indians of the Willamette Valley*, Mission Mill Museum Association, Inc., Salem, OR.

MacLennan, C. (1988), 'The Democratic Administration of Government', in M.V. Levine, C. MacLennan, J.J. Kushma, C. Noble, J. Faux M.G. Raskin, *The State and Democracy: Revitalizing America's Government,* Routledge, New York, pp. 49-78.

Marcy, A.D. (1993), *Identification of Probable Groundwater Flow Paths at an Inactive Uranium Mine Using Hydrochemical Models*, presented at 10th National Meeting of the American Society for Surface Mining and Reclamation, Spokane WA, May 16-19, 1993, also at Environmental Protection Agency, Site Listing Docket for the Midnite Mine Superfund Site, Seattle, WA.

Marcy, A.D., Scheibner, B.J., Toews, K.L. and Boldt, C.M.K. (1994), *Hydrogeology and Hydrochemistry of the Midnite Mine, Northeastern Washington*, US Department of the Interior, Bureau of Mines, Washington, DC and Environmental Protection Agency, Site Listing Docket for the Midnite Mine Superfund Site, Seattle, WA.

May, D.L. (1994), *Three Frontiers: Family, Land and Society in the American West, 1850-1900*, Cambridge University Press, Cambridge, UK.

McCool, S.F. and Guthrie, K. (2001), 'Mapping the Dimensions of Successful Public Participation in Messy Natural Resources Management Situations', *Society and Natural Resources*, Vol. 14, pp. 309-323.

McGee, T.K. (1999), 'Private Responses and Individual Action: Community Responses to Chronic Environmental Lead Contamination', *Environment and Behavior*, Vol. 31, pp. 66-83.

Mencken, F.C. (2000), 'Federal Spending and Economic Growth in Appalachian Counties', *Rural Sociology*, Vol. 65, pp. 126-147.

Merkhofer, M.W. (1987), *Decision Science and Social Risk Assessment: A Comparative Evaluation of Cost-benefit Analysis, Decision Analysis, and Other Formal Decision Aiding Approaches*, D. Reidel Publishing Company, Boston, MA.

Mohai, P. (1995), 'The Demographics of Dumping Revisited: Examining the Impact of Alternative Methodologies in Environmental Justice Research', *Virginia Environmental Law Journal*, Vol. 14, pp. 615-653.

Moon, J.D. (1993), *Constructing Community: Moral Pluralism and Tragic Conflict*, Princeton University Press, Princeton, NJ.

Mullen, F.C. (1971), *The Land of Linn,* Dalton's Printing, Lebanon, OR.

National Environmental Justice Advisory Council (15 November 2000), *Guide on Consultation and Collaboration with Indian Tribal Governments and the Public Participation of Indigenous Groups and Tribal Members in Environmental Decision Making*, a project of a work group of the Indigenous Peoples Subcommittee of the National Environmental Justice Advisory Council, a federal advisory committee to the United States Environmental Protection Agency, Washington DC.

National Parks Service (2002), 'What was the Homestead Act?', found at http://www.nps.gov/home/ homestead_act.htm, 23 July 2002.

Nelkin, D. and Brown, M.S. (1994), *Workers at Risks: Voices from the Workplace,* University of Chicago Press, Chicago.

Newcomb, M.D. (1986), 'Nuclear Attitudes and Reactions: Associations with Depression, Drug Use, and Quality of Life', *Journal of Personality and Social Psychology*, Vol. 50, pp. 906-920.

Nichols, D.W. and Scholz, A.T. (1987). *Blue Creek Metals Analysis: Concentrations of Al, Cd, Mn, U, Sr, Zn, and Ni in Whole Eviscerated Fish, Fish Livers, Invertebrates and Water,* Upper Columbia United Tribes Fisheries Center, Technical Report No. 9, Environmental Protection Agency, Site Listing Docket for the Midnite Mine Superfund Site, Seattle, WA.

Nisbet, R. (1990), *The Quest for Community: A Study in the Ethics of Order and Freedom*, ICS Press, San Francisco, CA.

Occupational Safety and Health Administration (1 October 2002), 'Health Guidelines for Zirconium and Zirconium Compounds', found at http://www.osha-slc.gov/SLTC/health guidelines/zirconiumandcompounds/recognition.html#healthhazard, 13 June 2003.

Oregon Legislative Commission on Indian Services (1999), '1999-01 Oregon Directory of American Indian Resources', found at http://www.leg.state.or.us/cis/directory99. pdf, 23 July 2002.

Oregon State Health Division, Radiation Control Section (July 1977a), *Preliminary Report: Radiological Aspects of Wah Chang Operations*, Portland, OR

Oregon State Health Division (August 1977b), 'Wah-Ching, Wah Chang', *Oregon Health*, Vol. 55(8), pp. 1-4.

Oregonian (unknown author) (1983), 'Wah Chang Fined $4,000 for Fire', *The Oregonian*, 8/23/1983.

Ostry, A.S., Hertzman, C. and Teschke, K. (1995), 'Community Risk Perception and Waste Management: A Comparison of Three Communities', *Archives of Environmental Health*, Vol. 50(2), pp. 95-102.

Park, R., Burgess, E. and McKenzie, R. (1925), *The City*, University of Chicago Press, Chicago.

Parsons, T. (1951), *The Social System*, Free Press, Glencoe, IL.

Pellow, D.N. (2002), *Garbage Wars: The Struggle for Environmental Justice in Chicago*, MIT Press, Cambridge, MA.

Pellow, D.N. and Sun-Hee Park, L. (2002), *The Silicon Valley of Dreams: Environmental Injustice, Immigrant Workers, and the High-tech Global Economy*, New York University Press, New York.

Perrow, C. (1984), *Normal Accidents: Living with High Risk Technologies*, Basic Books, New York.

Perrow, C. (1994), 'Accidents in High-risk Systems', *Technology Studies*, Vol. 1, pp. 1-38.

Petersen, K.C. (1987), *Company Town: Potlatch, Idaho, and the Potlatch Lumber Company*, Washington State University Press, Pullman, WA.

Pevar, S.L. (1992), *The Rights of Indians and Tribes: The Basic ACLU Guide to Indian and Tribal Rights* (second edition), Southern Illinois University Press, Carbondale, IL.

Pfeffer, M.J., Schelhas, J.W. and Day, L.A. (2002), 'Forest Conservation, Value Conflict, and Interest Formation in a Honduran National Park', *Rural Sociology*, Vol. 66, pp. 382-402.

Prucha, F.P. (1962), *American Policy in the Formative Years: The Indian Trade and Intercourse Acts 1790-1984*, Harvard University Press, Cambridge, MA.

Putnam, L.L. and Wondolleck, J.M. (2003), 'Intractability: Definitions, Dimensions, and Distinctions', in R.J. Lewicki, B. Gray and M. Elliot (eds), *Making Sense of Intractable Environmental Conflicts: Concepts and Cases*, Island Press, Washington, DC, pp. 35-59.

Quarantelli. E.L. (ed.) (1998), *What is a disaster?*, Routledge, New York.

Rawhide Press, (November 1984) 'Unemployment Statistics, Spokane Indian Reservation', *Rawhide Press*, Wellpinit WA.

Redfield, R. (1947), 'The Folk Society', *American Journal of Sociology*, Vol. 52, pp. 293-308.

Reich, M.R. (1983), 'Environmental Politics and Science: The Case of PBB Contamination in Michigan', *American Journal of Public Health*, Vol. 73(3), pp. 302-313.

Renn, O., Burns, W.J, Kasperson, J.X., Kasperson, R.E. and Slovic, P. (1992), 'The Social Amplification of Risk: Theoretical Foundations and Empirical Application', *Journal of Social Issues*, Vol. 48, pp. 137-160.

Rifkin, J. (1995), *The End of Work: The Decline of the Global Labor Force and the Dawn of the Post-market Era*, Putnam, New York.

Rosa, E.A. (1998), 'Metatheoretical Foundations of Post-normal Risk', *Journal of Risk Research*, Vol. 1, pp. 15-44.

Rosa, E.A. and Clark, D.L., Jr. (1999), 'Historical Routes to Technological Gridlock: Nuclear Technology as Prototypical Vehicle', *Social Problems and Public Policy*, Vol. 7, pp. 21-57.

Rosa, E.A. and Dunlap, R.E. (1994), 'Nuclear Power: Three Decades of Public Opinion', *Public Opinion Quarterly*, Vol. 58, pp. 295-325.

Rowan, K.E. (1991), 'Why Rules for Risk Communication are not Enough: A Problem-solving Approach to Risk Communication', *Risk Analysis*, Vol. 14, pp. 365-374.

Rubin, I. (1977), 'Function and Structure of Community: Conceptual and Theoretical Analysis', in R.L. Warren (ed.), *New Perspectives on the American Community* (third edition), Rand McNally College Publishing Company, Chicago, pp. 108-118.

Ruby, R.H. and Brown J.A. (1970), *The Spokane Indians: Children of the Sun*, University of Oklahoma Press, Norman, OK.

Rundmo, T. (1995), 'Perceived Risk, Safety Status, and Job Stress Among Injured and Noninjured Employees on Offshore Petroleum Installations', *Journal of Safety Research*, Vol. 26, pp. 87-97.

Sagan, S.D. (1993), *The Limits of Safety: Organizations, Accidents, and Nuclear Weapons*, Princeton University Press, Princeton, NJ.

Satterfield, T., Slovic, P., Gregory, R., Flynn, J. and Mertz, C.K. (2000), *Risk Lived, Stigma Experienced*, Decision Research, Eugene, OR.

Scholz, A., Peone T., Uehara, J., Geist, D. and Barber, M. (March 1988), *Rainbow Trout Population Estimates in Blue Creek, Spokane Indian Reservation, from 1985 to 1987: Detecting Impacts of Uranium Mine Discharge on the Rainbow Trout Population*, Upper Columbia United Tribes Fisheries Center, Fisheries Technical Report No. 10, Environmental Protection Agency, Site Listing Docket for the Midnite Mine Superfund Site, Seattle, WA.

Schuler, D. (1996), *New Community Networks: Wired for Change*, Addison Wesley Publishing Company, New York.

Schulz, D.R. (19 August 2001), 'Speaking to Survival', *Awakened Woman E-Magazine*, found at http://www.awakenedwoman.com/native_women.htm, 23 July 2002.

Scott, L.M. (1924), *History of the Oregon Country, Volume II*, The Riverside Press, Cambridge, MA.

Selznick, P. (1992), *The Moral Commonwealth: Social Theory and the Promise of Community*, University of California Press, Berkeley, CA.

Sharp, J.S. (2002), 'Locating the Community Field: A Study of Interorganizational Network Structure and Capacity for Community Action', *Rural Sociology*, Vol. 66, pp. 403-424.

Shepherd Miller, Inc. (12 July 1991), *Midnite Mine Reclamation Plan* prepared for Dawn Mining Company, Environmental Protection Agency, Site Listing Docket for the Midnite Mine Superfund Site, Seattle, WA.

Short, J.F., Jr. (1984), 'The Social Fabric at Risk: Toward the Social Transformation of Risk Analysis', *American Sociological Review*, Vol. 49, pp. 711-725.

Short, J.F., Jr. and Clarke, L. (eds) (1992), *Organizational, Uncertainties, and Risk*, Westview Press, Boulder, CO.

Shrader-Frechette, K. (2002), *Environmental Justice: Creating Equality, Reclaiming Democracy*, Oxford University Press, Oxford.

Shriver, T.E., Cable, S., Norris, L. and Hastings, D.W. (2000), 'The Role of Collective Identity in Inhibiting Mobilization: Solidarity and Suppression in Oak Ridge', *Sociological Spectrum*, Vol. 20, pp. 41-64.

Sjöberg, L. and Drtot-Sjöberg, B.M. (1991), 'Knowledge and Risk Perceptions Among Nuclear Power Plant Employees', *Risk Analysis*, Vol. 11, pp. 607-618.

Slezak, P. (1994), 'The Social Construction of Social Constructionism', *Inquiry*, Vol. 37, pp. 139-157.

Slovic, P. (1992), 'Perception of Risk: Reflections on the Psychometric', in S. Krimsky and D. Golding (eds), *Social Theories of Risk*, Praeger, London, pp. 117-152.

Slovic, P. (1997), 'Trust, Emotion, Sex, Politics and Science: Surveying the Risk-assessment Battlefield', in M.H. Bazerman, D.M. Messick, A.E. Tenbrunsel and K.A. Wade-Benzoni (eds), *Environment, Ethics and Behavior,* New Lexington, San Francisco, pp. 277-313.

Smith, P.D. and McDonough, M.H. (2001), 'Beyond Public Participation: Fairness in Natural Resource Decision Making', *Society and Natural Resources,* Vol. 14, pp. 239-249.

Sokolowska, J. and Tyszka, T. (1995), 'Perception and Acceptance of Technological and Environmental Risks: Why are Poor Countries Less Concerned?', *Risk Analysis,* Vol. 15, pp. 733-743.

Spokane Tribe of Indians (9 September 1999), 'Spokane Tribal Profile', found at http://www.npaihb.org/profiles/spokane.html, 25 June 2002.

Stanard, E.E. (25 August 1948), 'Last of the Calapooia Tribe', *Democrat-Herald,* Albany, OR.

Stein, M.R. (1960), *The Eclipse of Community: An Interpretation of American Studies,* Princeton University Press, Princeton, NJ.

Steingraber, S. (1998), *Living Downstream: A Scientist's Personal Investigation of Cancer and the Environment,* Vintage Books, New York.

Stern, P.C. (1993), 'A Second Environmental Science: Human-environment Interaction', *Science,* Vol. 26(June 25), pp. 1897-1899.

Stern, P.C. and Dietz, T. (1994), 'The Value Basis for Environmental Concern', *Journal of Social Issues,* Vol. 50, pp. 65-84.

Stevens County Rural Development Planning Council (1961), *Overall Rural Area Economic Development Program,* Stevens County, Colville, WA.

Sumioka, S.S. and Dion, N.P. (1983), *Water Quality Conditions Near Midnite Mine, Stevens County, Washington,* U.S. Geological Survey Water Resources Investigations Report 83 Document No. 3957, Environmental Protection Agency, Site Listing Docket for the Midnite Mine Superfund Site, Seattle, WA.

Sumioka, S.S. (1991), *Quality of Water in an Inactive Uranium Mine and its Effects on the Quality of Water in Blue Creek, Stevens County, Washington, 1984-85,* US Geological Survey Water Resources Investigations Report 89-4110, Environmental Protection Agency, Site Listing Docket for the Midnite Mine Superfund Site, Seattle, WA.

Superfund Technical Assessment and Response Team (START) (September 1998), *Midnite Mine Expanded Site Inspection Report,* TDD 97-07-0010, Environmental Protection Agency, Site Listing Docket for the Midnite Mine Superfund Site, Seattle, WA.

Swanson, L.E. (2002), 'Rural Policy and Direct Local Participation: Democracy, Inclusiveness, Collective Agency, and Locality-based Policy', *Rural Sociology,* Vol. 66, pp. 1-21.

Swearengen, J.C. (1996), 'Brownfields and Greenfields: An Ethical Perspective on Land Use', *Environmental Ethics,* Vol. 21, pp. 277-292.

Tauxe, C.S. (1995), 'Marginalizing Public Participation in Local Planning: An Ethnographic Account', *Journal of American Planning Association,* Vol. 61, pp. 471-480.

Taylor, H.C., Jr. and Hoaglin, L.L. (1962), 'The "Intermittent Fever" Epidemic of the 1830's on the Lower Columbia River', *Ethnohistory,* Vol, 9(2), pp. 160-178.

Töennies, F., translated by C. Loomis (1940), *Fundamental Concepts of Sociology (Gemeinschaft and Gesellschaft),* American Book Company, New York.

Tolbert, C.M., Irwin, M.D., Lyson, T.A. and Nucci, A.R. (2002), 'Civic Community in Small-town America: How Civic Welfare is Influenced by Local Capitalism and Civic Engagement', *Rural Sociology,* Vol. 67, pp. 90-113.

Tyler, S.L. (1973), *A History of Indian Policy*, Government Printing Office, Washington DC.

U.S. Bureau of Mines (6 June 1994), *Midnite Mine Characterization Work Plan, Final Draft*, US Department of the Interior, Washington, DC, also at Environmental Protection Agency, Site Listing Docket for the Midnite Mine Superfund Site, Seattle, WA.

U.S. Geological Survey (prepared in cooperation with U.S. Bureau of Mines) (1996), *Inventory, Characterization, and Water Quality of Springs, Seeps, and Streams Near Midnite Mine, Stevens County, Washington*, Open-file Report 96-115, U.S. Department of Interior, Tacoma, WA, also at Environmental Protection Agency, Site Listing Docket for the Midnite Mine Superfund Site, Seattle, WA.

URS Corporation (20 September 2000), *Ecological Characterization of Midnite Mine*, Project No. 53F4001800.07, Environmental Protection Agency, Site Listing Docket for the Midnite Mine Superfund Site, Seattle, WA.

URS Corporation (30 October 2001a), *Draft Work Plan: Human Health Risk Assessment Work Plan for the Midnite Mine Superfund Site*, Project No. 53F4001800.00, Environmental Protection Agency, Site Listing Docket for the Midnite Mine Superfund Site, Seattle, WA.

URS Corporation (20 December 2001b), Technical memorandum to Ellie Hale from Jim Scott concerning Midnite Mine, Detail Area 4, Project No. 53F4001800.01 Task 18110, Environmental Protection Agency, Site Listing Docket for the Midnite Mine Superfund Site, Seattle, WA.

Vaughan, E. (1995), 'The Socioeconomic Context of Exposure and Response to Environmental Risk', *Environment and Behavior*, Vol. 27, pp. 454-489.

Victor, F.F. (1872), *All Over Oregon and Washington: Observations on the Country, Its Scenery, Soil, Climate, Resources and Improvements*, John H. Carmany and Co., San Francisco.

Vicusi, W.K. (1983), *Risk by Choice: Regulating Health and Safety in the Workplace*, Harvard University Press, Cambridge, MA.

Vorkinn, M. and Riese, H. (2001), 'Environmental Concern in a Logical Context: The Significance of Place Attachment', *Environment and Behavior*, Vol. 33, pp. 249-263.

Walter, J. (12 April 1989), 'Uranium Levels Threaten an Idyllic Spot', *Spokesman Review*.

Warren, R. (1977), 'The Good Community', in R.L. Warren (ed.), *New Perspectives on the American Community* (third edition), Rand McNally College Publishing Company, Chicago, pp. 535-545.

Warren, R. (1978), *The Community in America*, Rand McNally College Publishing Company, Chicago.

Washington Department of Social and Health Services (18 July 1980), *Revised Draft Environmental Impact Statement: Dawn Mining Company Mill Tailings Expansion Project*, Washington Department of Social and Health Services, Division of Radiation Protection, Olympia, WA.

Washington State Department of Social and Health Services (6 February 1989), *Draft Environmental Impact Statement: Proposed Closure of the Dawn Mining Company Uranium Millsite in Ford Washington*, Washington Department of Social and Health Services, Division of Radiation Protection, Olympia, WA.

Washington State Department of Health (1991), *Final Environmental Impact Statement for Closure of the Dawn Mining Company Uranium Millsite*, Washington State Department of Health, Division of Radiation Protection, Olympia, WA.

Washington State Department of Health (15 July 1994), *Draft Supplemental Environmental Impact Statement: Closure of the Dawn Mining Company Uranium Millsite in Ford Washington*, Washington State Department of Health, Division of Radiation Protection, Olympia, WA.

Washington State Department of Health (9 March 2001), news release entitled 'Department of Health's Release of Sherwood Uranium Mill Site License Marks End of Nine-year Reclamation Project; Is First in U.S. to Receive License Termination Approval by NRC', Washington State Department of Health, Division of Radiation Protection, Olympia, WA.

Weart, S.R. (1988), *Nuclear Fear: A History of Images*, Harvard University Press, Cambridge, MA.

Weber, M. (1978), *Economy and Society*, University of California Press, Berkeley, CA.

Weber, M., translated by D. Martindale and G. Neuwirth (1958), *The City*, The Free Press, New York.

Weller, S.C. and Kimball Romney, A. (1988), 'Systematic Data Collection', *Qualitative Research Methods Series*, Volume 10, Sage Publications, Inc., Newbury Park, CA.

Wellpinit School District (2002), 'A Short History of the Spokane', found at http://www.wellpinit.wednet.edu/spokan/history/shorhist.php, 18 June 2002.

White, R. (1991), *'It's Your Misfortune and None of My Own': A New History of the American West*, University of Oklahoma Press, Norman, OK.

Whiteman, D.C., Dunne, M.P. and Burnett, P.C. (1995), 'Psychological and Social Correlates of Attrition in a Longitudinal Study of Hazardous Waste Exposure', *Archives of Environmental Health*, Vol. 50, pp. 281-286.

Wilkinson, K.P. (1991), *The Community in Rural America*, Greenwood, New York.

Williams, W.L., Jr. (2002), *Determining Our Environments: The Role of Department of Energy Citizen Advisory Boards*, Praeger, London.

Wing, S. (1994), 'The Limits of Epidemiology', *Medicine and Global Survival*, Vol. 1(2), pp. 74-86.

Winner, L. (1993), 'Upon Opening the Black Box and Finding it Empty: Social Constructivism and the Philosophy of Technology', *Science, Technology and Human Values*, Vol. 18, pp. 362-378.

Williams, W.L., Jr. (2002), *Determining Our Environments: The Role of Department of Energy Citizen Advisory Boards*, Praeger, London.

Wood, G.S., Jr., and Judikis, J.C. (2002), *Conversation on Community Theory*, Purdue University Press, West Lafayette, IN.

Woodward, W.L. (1971), 'A Report of an Inventory and Study, Water Resources and Utilities', consulting paper prepared for the Spokane Tribe of Indians.

Wynccoop, D.C. (1969), *Children of the Sun: A History of the Spokane Indians*, Comet and Cole, Spokane, WA.

Wynne, B. (1992), 'Risk and Social Learning: Reification to Engagement', in S. Krimsky and D. Golding (eds), *Social Theories of Risk*, Praeger, London, pp. 275-297.

Yen, I.H. and Syme, S.L. (1999), 'The Social Environment and Health: A Discussion of the Epidemiologic Literature', *Annual Reviews in Public Health*, Vol. 20, pp. 287-308.

Zavestoski, S., Mignano, F., Agnello, K., Darroch, F. and Abrams, K. (2002), 'Toxicity and Complicity: Explaining Consensual Community Response to a Chronic Technological Disaster', *The Sociological Quarterly*, Vol. 43, pp. 385-406.

Index

Anderson 6, 17, 43, 82
Axelrod 6, 9

Beck 1, 7, 10, 13, 17, 24, 34
Boholm 9
Brown 13, 130
Bullard 1

Carson 6-7, 34
Case selection 25-26
CERCLA
    see Superfund
Clark 11, 17, 20, 132
Clarke 9, 11, 14, 15
Chess 2, 11, 13, 106, 110
Cohen 8, 41
Colonization acts
    Donation Act of 1850 48, 51, 151-2
    General Allotment Act of 1887 50,
        155-6
    Homestead Act of 1862 49, 153-4
    Indian Removal Act of 1830 48, 149
    Indian Reorganization Act of 1934
        50, 156
    Land Ordinance of 1785 148-9
    Preemption Act of 1841 48, 150-1
    Termination Act of 1954 50, 156
Communitarian theory 5, 12-13, 15, 18
Community
    cohesion 6, 7, 18, 19
    identity 8-11, 126-30
    involvement policies 3, 19-22, 38-9,
        132-33, 138-40
Conflict 2, 16, 19-20, 82134
Couch 8, 10, 11, 12-13, 15-16

Data
    analysis methods 32
    collection 29-32

Dawn Mining Company 26-8, 54, 56-7,
    61-2, 63, 65, 85-109, 112, 117-19,
    121, 124, 128-9, 130-31, 134-37
Dawn Watch 106, 115, 131, 135
Dietz 7, 9
Disaster research 7, 10, 12-16
Dombrowsky 13
Donation Act of 1850 48, 51, 151-2
Douglas 1, 16
Dunlap 9
Durkheim 17-18

Environmental justice 1, 103, 115, 131
Epidemiology 2
Erikson 7, 10, 13, 14, 15
Etizioni 6, 18, 110

Flynn 9, 14, 17
Freie 12
Frey 9, 11
Freundenburg 6, 17
Fur trading 46-7, 145-6

General Allotment Act of 1887 50,
    155-6
Gilbert 16
Goffman 10
Golding 9
Grand Ronde Reservation 48, 50, 152-3,
    156
Gray 2, 16, 83, 132
Gridlock 11, 17, 132

Hanford 86, 96
Hannigan 6, 8, 9, 16, 17, 34, 41
Hazard Ranking System 36-8, 71, 93-5
Hewitt 13, 15
Homestead Act of 1862 49, 153-4
Human ecology 5

Indian Removal Act of 1830 48, 149
Indian Reorganization Act of 1934 50,
        155
Indigenous people
        Spokane Tribe 42, 45, 48-9, 51,
                52-3, 63, 64-5, 144-5, 151-2,
                155-7
        Willamette Valley, 42, 44, 48, 50-1,
                52, 142-5, 149, 151-3, 156-7
Industrial interest 51-64

Kasperson 9
Kaufman 8, 41
Kemmis 6, 12
Krimsky 9
Kroll-Smith 8, 10, 11, 12, 15-16

Land Ordinance of 1785 148-9
Land use beliefs 51-64
Local Citizens Monitoring Committee
        103
Love Canal 38, 136

Manhattan Project 86
Marx 17-18
Methods24-32
Missionaries 47, 147-8

National Contingency Program 35
Networks 6, 12, 15
NEPA 2, 3, 22
NIMBY 62, 85
Nisbet 18, 110
NPL 3, 25-26

Operational premises 28-9, 133-36

Parsons 12
Perrow 13
Place attachment 5, 18
Power 12-17, 108, 111-14, 116, 130-32,
        139-40
Preemption Act of 1841 48, 150-1
Preliminary assessment petitions 35-6
Public participation 2, 3, 11, 13, 15,
        17-20, 38-9, 80-84, 100-107, 110-25,
        132-33, 138-40

Quartanelli 7, 14

Railroad 154-5
RCRA 22
Renn 9
Rifkin 18, 110
Risk
        communication 2, 11, 13, 19, 20,
                76-9, 81-84, 108-109, 133, 134
        odors 57
        industrial wastes 55, 70-71, 72-8,
                89-90, 99
        perceptions 9-10, 72-8, 107
        radiation 59, 72-8, 90-92, 95-100,
                106
        research approaches 4, 21, 22
Rosa 6, 9, 10, 11, 17, 20, 34, 132
Rural-urban dichotomy 5, 6

Sagan 13
Selznick 5, 6, 8, 12, 15, 18, 41, 110
Shared history 5-8, 126-30
Sherwood Mine and Mill 27, 57, 59, 63,
        106, 128
Short 9-10, 14, 15
Shrader-Frechette 131
Slovic 9, 11,14
Social
        amplification of risk framework
                9, 21, 34, 139-40
        constructionism 5, 6, 8, 9, 10, 34
        interaction 5-7, 12, 18
Spokane reservation 49-50, 59, 88, 154
Stern 9
Stigma 10-11
Superfund 2, 3, 6, 10, 22-3, 34-40

Technical Assistance Grants 39
Teledyne Wah Chang Albany 26, 33, 56,
        58, 60, 61, 62-3, 64, 65, 67-84, 111,
        114, 116-7, 119-20, 121, 124, 127-9,
        130-31, 133-37
Termination Act of 1954 50, 156
Times Beach 38
Tonnies 18
Trust 11, 17, 20, 61-2, 106, 108, 114,
        131, 132, 134

Uranium
    exploration programs 86
    yellow cake 56, 86

Warren 6, 12, 15
Weart 9-10
Weber 17-18
Wildavsky 1, 16
Wilkinson 6
Wing 2
Wynne 7

Zirconium 54, 67-8